HOW IRISH SCIENTISTS CHANGED THE WORLD

This book is dedicated to the memory of my creative and fun-loving mother Fiona Duke (née Herr), 1939-2007. You always said that I had a book in me. Well, finally, I have proved you right. We miss you.

HOW IRISH SCIENTISTS CHANGED THE WORLD

Seán Duke

LONDUBH BOOKS

First published in 2013
by Londubh Books
18 Casimir Avenue, Harold's Cross, Dublin 6w, Ireland
www.londubh.ie
1 3 5 4 2
Origination by Londubh Books; cover by bluett
Printed by ScandBook AB, Falun Sweden
ISBN: 978-1-907535-25-3

CONTENTS

(Denotes Nobel prizewinner)*

ACKNOWLEDGEMENTS

I would like to thank everyone who provided me with information or pointed me in the direction of relevant information while I was researching and writing this book.

The assistance of Professor Philip Walton, retired Professor of Physics, NUI Galway, in building a picture of his father, Ernest Walton, was precious. Philip, as a physicist himself, was able to explain the science behind his father's work and its importance in the birth of the atomic age. Philip was kind enough to relate many interesting and sometimes amusing anecdotes about his dad, as a scientist and father.

Thanks too to Dr Eric Finch, Department of Physics, TCD, the last man to be hired by Professor Walton before he retired in the 1970s. Dr Finch was a mine of information on all things to do with his former boss and his affection for him as a person was clear.

Belfast-born Jocelyn Bell Burnell is the only living scientist portrayed in this book and one of only two women. Now aged seventy, she remains an active researcher based in the University of Oxford and a strong campaigner for more women becoming full-time scientists. She is a superb role model for young Irish women scientists and young Irish scientists in general and, like many top scientists, she is an engaging and entertaining public speaker. Thank you, Jocelyn, for giving me your time on more than one occasion, to help with this book.

Thanks too, to Luke Drury of the Dublin Institute for Advanced Studies, for helping with my research into the life and work of Erwin Schrödinger while he was in Dublin (1940-57).

Otto Glaser, a businessman now living in Howth, is one of the few people who can clearly recall Schrödinger's life in Dublin, both as a scientist and a socialite. Like Schrödinger, Glaser fled his native

Austria and Nazi rule during the Second World War and ended up in Ireland. He met Schrödinger at the Irish Austrian Society in Dublin while he was still a schoolboy. Thanks to him for spending several hours with me talking about Schrödinger.

William Rowan Hamilton was one of our greatest mathematicians and I would like to thank a modern mathematician, Fiacre O'Cairbre, for his insights into Hamilton and for giving so readily of his time. Fiacre, who is based in NUI Maynooth, organises the annual Hamilton walk along the Royal Canal in October, to commemorate the location where Hamilton made a major discovery.

I'd like to thank Matthew Trainer of Glasgow University's School of Physics and Astronomy for bringing William Thomson/Lord Kelvin to life and showing me many of his scientific instruments and his fine home.

Niall McKeith, curator of the National Science Museum in Maynooth, kindly directed me to texts that shed light on the work and character of Fr Nicholas Callan. Niall told me about the man, his sense of humour and his determination to conduct his electrical research, despite the apathy and even antipathy of his priestly colleagues.

Robert Mallett, a 19th-century Dubliner, was a talented engineer but his great legacy is to geology. On Killiney strand, in 1849, he showed, for the first time ever, that an earthquake, as represented by the explosion of gunpowder, is caused when rocks move and send out two types of energy waves. Tom Blake, of the geophysics unit at the Dublin Institute for Advanced Studies, set up the school's seismology programme. As part of his efforts to encourage more interest in seismology and science in general, Tom has been telling students and others about Robert Mallett. Thank you, Tom, for bringing Mallett to life.

I'd like to thank David McConnell, Professor of Genetics in TCD, for letting me know that Maurice Wilkins, who shared the 1962 Nobel prize with Francis Crick and James Watson for determining the structure of DNA, the genetic code, was – genetically – Irish. Maurice's parents, Edgar and Eveline, were Dubliners and his grandfather, William Wilkins, was headmaster at Dublin's High School (now in Rathgar). I would like to thank Alan Phelan, archivist at the High School, for his help.

The Irishness of Guglielmo Marconi might come as a surprise to some, particularly Italians, who proudly claim him as one of their greatest sons. The fact is, however, that Marconi was as Irish as he was Italian, by virtue of his mother, Annie Jameson. I would like to thank Peter Kennedy, Professor of Electrical and Electronic Engineering in UCC, for describing Marconi's work and his Irish connections.

I'd like to thank John Joyce, retired physics teacher and part-time tour guide at Birr Castle for his insights into the castle's two most famous residents, William and Charles Parsons. Thank you too to Garry Lyons, Adjunct Professor of Mechanical Engineering in TCD, for his insights into the life and work of Charles Parsons, for whom the Parsons Building in TCD is named.

My understanding of the self-taught mathematical genius who was George Boole was greatly enhanced by talking to Dr Gary McGuire, mathematician at UCD's CASL (Complex and Adaptive Systems Laboratory).

Thanks too, to the staff at the Clare County Library who helped point me in the right direction when I was investigating the life and times of Liscannor's submarine hero, John Holland.

Eoin Gill of Waterford Institute of Technology, one of the drivers of the Robert Boyle Summer School, held annually in Lismore Castle, was helpful on subject of Boyle's life. So was Michael Hunter, the author of *Boyle: Between God and Science* (2010), which explores Boyle's role in popularising the experimental method, among other things.

Norman McMillan, formerly a senior lecturer in the Institute of Technology in Carlow, provided insights into John Tyndall, the legendary Victorian scientist and popular author. Thanks to Norman and good luck to him in his efforts to reintroduce Tyndall to a new generation of Irish people, who might judge him for his science and not his politics.

Sincere apologies if I have forgotten anyone who helped me in any way while I was researching and writing this book.

Finally, I'd like to thank my publisher, Jo O'Donoghue, for her support and encouragement throughout this project and for her patience when the work schedule got 'off track'.

FOREWORD BY PATRICK J. PRENDERGAST, PROVOST, TRINITY COLLEGE DUBLIN

What good reason does Seán Duke have for putting the scientists in this volume together? 'They were all born in Ireland,' I hear you say, '…or worked in Ireland.' But science is not something that recognises national borders. Even in the 17th century, in the days of the great Robert Boyle, correspondence flowed between Europe's learned academies and nowadays information travels in seconds around the world.

How Irish Scientists Changed the World will show you why it makes sense to look at Irish scientists as a group. It has little to do with being Irish as such. 'No man is an island,' as John Donne said, and Irish scientists are linked in a multiplicity of ways: often by education; often too by family connections. We share our eureka moments with those who are nearest to us. They inspire us and we, in turn, sometimes inspire them.

The diversity of Irish science is striking. All manner of Irishness is represented and celebrated. Some did all their work in Ireland, like Fr Nicholas Callan of Maynooth, some emigrated to the United States to find fame, like the inventor of the submarine, John Holland. Yet others were immigrants finding inspiration in Ireland for discoveries that might never have been made elsewhere, such as Erwin Schrödinger, whose work in Dublin presaged the discovery of the genetic code, or George Boole in Cork, who invented the mathematics of computing. Jocelyn Bell Burnell from Northern Ireland, the only living scientist in this book, discovered pulsars when working in the University of Cambridge. I am pleased to say that she was recently elected a Pro-Chancellor of the University of Dublin. One of my favourite scientists in the book is Marconi, whose mother was from Enniscorthy. He conducted many radio experiments on visits to his mother's people in Ireland, so he is claimed for Irish science in this book. But an exclusive claim would be wrong because scientific achievements recognise no national borders: science stands to benefit everyone.

Scientific endeavour is driven by our innate human curiosity and changes how we look at the world. It can help to lift the fog of superstition and ignorance. Equally important are the opportunities science holds for enhancing our lives with better medicines, labour-saving machines and improved technologies of communication.

However, knowledge cannot advance by science alone. The great advances by Irish scientists were made at the same time as well-known cultural achievements in literature and the performing arts. Creative people are an inspiration for one another, be they poets or physicists – and some people are both! A sense of place also matters: the Nobel prizewinners Ernest Walton and Samuel Beckett were students together in Trinity College Dublin. Indeed, many of the scientists and engineers written about here are associated with TCD, as well as with the island of Ireland. Why a place can become a source of some of the world's greatest creativity is a mystery, a mystery that is partly unravelled in this wonderful book which I recommend wholeheartedly to a wide readership.

Patrick J Prendergast
September 2013

INTRODUCTION

Ireland is famed for its writers, musicians and actors; yet the huge impact of its scientists from the 17th century right up to today remains a largely untold story. This book tells that story by focusing on the lives and work of seventeen scientists, fifteen men and two women – just one of whom is still living – whose brilliance changed the world.

At first glance the case for Irish genius in science is weak. Ireland has just one science Nobel laureate – Ernest Walton, the atom-splitter – compared to four in literature: William Butler Yeats, George Bernard Shaw, Samuel Beckett and Seamus Heaney. The imbalance is often cited as proof that we are a nation of artists and dreamers. This is a clichéd and inaccurate view but one that we as Irish people are comfortable with, as it recalls our great early-Christian period when we kept European civilisation alive in the Dark Ages. In those idyllic pre-colonisation days there were scribes and musicians but no scientists.

There is a sense that science and industry are British creations, not ours. Therein perhaps lies a clue to why Ireland has not honoured its greatest scientists. Science was seen as something imposed or imported. This was a view reinforced by the fact that the leading scientists were mostly Protestant members of the ruling class. Some of them, such as John Tyndall, one of the most famous scientists of the Victorian era, made enemies in the Catholic community; in Tyndall's case, this resulted from his stridently anti-Catholic, pro-Union letters to *The Irish Times.*

When independence came in 1922, Tyndall and people like him were on the wrong side of history. The new Irish state had no wish to laud the work of domestic enemies. Science in Ireland was an elitist business, available for the tiny number of people who could access third-level education. Science was, therefore, largely for Protestant Irishmen (there were exceptions of course) who attended Trinity College Dublin or one of the Queen's Colleges set up in the mid-19th century in Belfast, Galway and Cork. The National University of Ireland, as we now know it, did not come into being until 1908.

Earlier Catholic agitation resulted in the setting up of the Royal College of St Patrick in Maynooth in 1795 – for Catholics only. Significantly, however, the focus of St Patrick's soon became solely the education of Catholic priests and the teaching of science was not considered to be of any importance. This was why the pioneering electrical investigations of Nicholas Callan were met with apathy by his priestly colleagues in the 1830s and 1840s. The underlying message was that science was part of the invader's culture. Catholics needed their own learning and it should focus on theology and history, not science.

The students who entered TCD had no such political or religious baggage. Throughout the Victorian era the college produced talented scientists and engineers. Five TCD graduates feature in these pages, four of whom lived under Queen Victoria: William Parsons, who built the world's largest astronomical telescope at the height of the Great Famine; his son, Charles, the inventor of the steam turbine; Robert Mallett, who first identified the true scientific reasons why earthquakes occur; and William Rowan Hamilton, who is considered one of the greatest mathematicians of all time. In the 20th century, Ernest Walton joined this elite group. He it was who, with odds and ends, on a shoestring budget, built the apparatus that split the atom in 1932.

It will be noticeable from a cursory glance through this book that it includes very few women scientists. There are just two, both astronomers: Annie Maunder, who discovered the link between sunspots and the earth's climate; and Jocelyn Bell Burnell – the only living scientist featured – who discovered pulsars, a new type of star, in 1967. The reason for the lack of women is simple. This list is based on achievement and for centuries Irishwomen, north and south, Catholic and Protestant, were excluded from scientific training. This discrimination has been tackled seriously only in recent times. Annie Maunder was one of the first women astronomers to be paid for her work but had to take a post in the Royal Observatory in Greenwich at a level far below her ability.

More recently, Jocelyn Bell Burnell, born into a Quaker family in Belfast and reared in Lurgan, County Armagh, received a scientific education in the 1950s only when her parents insisted that the school provide it. My own mother, Fiona Duke, who was a botanist, entered University College Dublin in 1958 to study science without ever having studied a single science subject at school. This was typical of the time and explains why women did not achieve great things in science.

In this book I have boldly claimed, on behalf of Ireland, three additional scientific Nobel laureates: Maurice Wilkins, Guglielmo Marconi and Erwin Schrödinger. Let's look at Ireland's claims on Wilkins, Marconi and Schrödinger.

Maurice Wilkins won the Nobel Prize in Physiology or Medicine in 1962, along with James Watson and Francis Crick, for his work in discovering the structure of DNA. Two nations have claimed Wilkins as their own: England, where Wilkins was educated and did most of his work and New Zealand, where he was born. Yet, genetically speaking, Wilkins was 100 per cent Irish.

Ireland's claims to Marconi, the wireless wizard, are even stronger. His mother was Annie Jameson, a talented singer and a member of the Jameson distilling family of Wexford. It was his mother who helped the young Marconi to secure financial support when he really needed it. His first wife was Beatrice O'Brien, the daughter of the 14th Baron Inchiquin of Dromoland Castle, County Clare. Marconi set up his famous transatlantic wireless radio communication station in Clifden, County Galway, and achieved many other career milestones on Irish soil.

It might come as a surprise to some readers that Erwin Schrödinger, a founder of the science known as quantum mechanics and one of the true greats of modern science, was an Irish citizen. While lecturing in Berlin in the 1930s, Schrödinger ran into trouble with the Nazis by opposing their burning of books they didn't like and the promotion of Nazis in his university. He was forced to flee and ended up back in his native Austria. After the German *Anschluss* with Austria in 1938 he again had to leave his country in a hurry. Taoiseach Éamon de Valera, himself a mathematician, saw his chance and offered Schrödinger a job and sanctuary in Ireland. Schrödinger arrived in Dublin in 1940 to work as head of the new Dublin Institute for Advanced Studies. He stayed for seventeen years, lived in Clontarf and took out full Irish citizenship.

If we accept the inclusion of these four scientists this leaves the science-literature Nobel tally at four apiece. So much for Ireland being home only to world-class writing talent!

People who are familiar with the life and work of Jocelyn Bell Burnell will know that astronomers now agree that she too should have received a Nobel Prize in science. The discovery of pulsars was hers but two senior male scientists took the credit and won Nobel prizes for it. Bell Burnell was overlooked because she was a woman, only twenty-four and a post-doctoral researcher. If we include Bell Burnell in our list of Nobel laureates, the Irish laureate table tips over to 5:4 in favour of science over literature.

The west of Ireland is thinly represented in these pages but it has a giant in Liscannor's John Holland, who invented the first combat submarine commissioned by the US navy.

George Boole, inventor of the language of computers, was born in Lincoln but is included in this book because he spent his entire academic career and did his greatest work in Queen's College, Cork. There he invented the language of computers and electronic devices.

Another scientific colossus described here is Robert Boyle, a key figure in promoting the use of experiment and the evidence it yielded, as the best way to understand nature.

These days, science in Ireland has become increasingly linked to industry, as the government seeks to get a return, in terms of jobs, from its investment in science This is not an environment in which everyone is comfortable but one scientist who would have thrived in it was the Belfast born scientist-entrepreneur William Thomson. Lord Kelvin, as he later became, is associated with the Kelvin temperature scale but his greatest contribution to science was his work with the first transatlantic cable.

Arguably, the man who did most to make Britain 'Great' was not British but Irish. Francis Beaufort, born in Navan, played a huge role in empire-building by providing British ships with accurate marine maps and facilitating the safe export of British goods worldwide. He also helped to make seafaring a far safer exercise for all mankind.

It is time to reclaim our scientific heritage in Ireland and to put forward our forgotten greats as role models for the future generation of scientists. A mature Ireland, a multicultural Ireland, one that places science at the centre of its plans for our future, should not care about the religious or political backgrounds or views of these scientists. They are Irish and among our most extraordinary talents. That should be enough.

Seán Duke
June 2013

Note: I use the abbreviation TCD for Trinity College Dublin throughout the text.

Part I

MAPS, EARTHQUAKES, ELECTRICITY AND CLIMATE

1

WHEN BRITANNIA RULED:

Francis Beaufort (1774-1857)

Born: Navan, County Meath

In 1829, Francis Beaufort, a fifty-five-year-old Irishman, finally got the job his talent richly deserved when he was appointed Chief Hydrographer (marine map-maker) to the British Admiralty. Beaufort, whose grandfather arrived in Ireland in the 1740s after fleeing religious persecution in northern France, had joined the British navy at the age of fifteen, after just three years of formal education. This largely self-taught man built the obscure and somewhat irrelevant Hydrographic Office up into the world's leading marine map-making and meteorology centre. Under his leadership a select team of ships' captains, who were also trained marine surveyors, charted many of the world's uncharted coastlines, seas and harbours and gathered scientific data that advanced numerous fields of science, notably biology, astronomy, meteorology and geology. Beaufort popularised technologies that had became available in the 19th century to measure sea depths, wind strength, current and tidal power more accurately and he applied his superb general knowledge of science and mathematics to solving all kinds of practical problems. By the time he retired in 1855, the navy made no big decisions without first consulting Dr Beaufort.

In the short term Beaufort's work cemented British dominance at sea by providing its navy with an enormous information advantage over potential maritime rivals. British captains had more detailed maps and meteorological information to hand than their enemies. This also meant that, for the first time in naval history, fewer British ships were lost as a result of an ignorance of seabed geography or of meteorology than were sunk by the enemy. Under Beaufort's direction, the navy also systematically charted safer routes by which the British merchant fleet could transport goods throughout the Empire. This was key to Britain's rise to superpower status during its 'glorious' imperial age, 1815-1915. In the longer term, Beaufort's legacy helped make seafaring safer for mariners and for all mankind. His renowned maps were widely used well into the 20th century.

Despite Beaufort's many achievements, his name is primarily linked with the Beaufort Wind Scale. He drew up this scale on board the first vessel he captained, the HMS *Woolwich*, as he dined alone in mid-Atlantic one evening in 1805. Beaufort was searching for a more precise way to measure wind than the loose terminology, wide open to individual interpretation, that had been in widespread use since the early 1600s. The weakness of the older method of assessment, which was based on wind speed, was that wind speed could vary greatly depending on where an observer was standing on the ship: for example, on a windswept lookout post or a calm deck a hundred feet below. Beaufort felt that it would be more scientific to assess wind based on the force of its action on the sails. This action would be clear for all to see and the same no matter where the observer was located.

As is often the case in science, the person who came up with the idea for the new wind scale did not receive his due credit. The scale that Captain Beaufort personally decided to adopt from 1805 onwards was almost identical to one that had been devised at the end of the 17th century by Alexander Dalrymple, the first man to be appointed Chief Hydrographer to the Admiralty. Dalrymple was a Scotsman and a pioneer of marine map-making.

Dalrymple himself drew heavily on the wind work of John Smeaton, an English engineer, often described as 'the father of civil engineering'. In the mid 18th century, when Smeaton wasn't building bridges, roads and harbours, he devised a way of measuring wind speed in windmills. This meant that the windmill designers could more accurately judge how much sail was required to produce a certain amount of wind speed and thereby power in their mills. This improved windmill efficiency and helped industry.

The reason that the Dalrymple Scale became known as the Beaufort Wind Scale was that it was Beaufort who ensured that it came into widespread use. This finally happened when, as Chief Hydrographer to the Navy, he sent out a memorandum on 28 December 1838, stating that from that date on all naval sea captains were obliged to use his new scale when assessing the strength of the wind. In the early 1860s, the British merchant fleet followed suit and adopted the Beaufort Wind Scale.

Under the old system, the measurements of wind were highly subjective and wind observers often disagreed over the meaning of terms. Nobody was exactly sure, for example, how a 'fresh gale' might be different to a 'stiff gale'. The new scale brought in more precise terminology upon which everyone could agree, as well as an associated number. The number zero on the new scale denoted 'calm', while 6 described a 'fresh breeze', 11 a "hard gale' and 13, a 'storm'.

When the powerful British navy, the lords of the high seas, adopted the new scale, the navies of many other seafaring countries followed suit, including the Dutch and the Chinese. The numbers on the scale also corresponded to the effect of wind on the ship's anemometer, a device that had been around for centuries and was based on a simple cup design, which grabbed the wind and rotated with it. The first anemometer had been invented in 1490 by Leon Battista Alberti, an Italian

Renaissance man with many interests (incidentally an Irishman, Dublin-born Dr John Robinson, Director of the Armagh Observatory, invented a more accurate anemometer in 1846, one with four cups, that moved in a horizontal direction proportional to the wind speed).

The new Beaufort Scale worked very well and was used increasingly as the 19th century went on, helping to move seafaring into the modern, scientific era. Eventually a metric system of measuring wind became more widely used than the old imperial scale but in Britain and Ireland the Beaufort scale still remains in use for shipping forecasts. Met Éireann issues a small craft warning if winds of Beaufort force 6 – exceeding 22 knots – are expected up to ten nautical miles offshore. Storm-force warnings are issued if force 9 winds of 52 knots or more are predicted, while hurricane-force warnings will go out if winds of greater than 64 knots are expected.

However, it is indisputable that Beaufort's lasting legacy to the world was in the field of marine mapping. Few would argue that he was the greatest marine mapper of all time and until very recently mariners were using his 19th-century admiralty maps to navigate their way around Irish coasts. They were the most accurate available to mariners until Ireland's National Seabed Survey (1999-2005) completed its work of charting Ireland's bays, harbours, coastline and offshore waters, using the latest technology available.

The Beaufort family arrived in Ireland when Beaufort's grandfather, Daniel de Beaufort, was offered the benefice of the Huguenot church in Navan, which also served Dublin's significant Huguenot population. Like many Huguenots before him, he had fled religious persecution in 18th-century France and sought sanctuary in Protestant England. After arriving in London, Daniel de Beaufort was minister at several London churches. One of his parishioners happened to be William Stanhope,

Earl of Harrington, and when the Earl was appointed Viceroy in Ireland in 1746 he brought Daniel to Ireland as his chaplain. (Daniel had anglicised the family name some time after his arrival in England by dropping the 'de' in de Beaufort.)

Daniel Beaufort senior served as rector in Navan for eighteen years until, in 1764, the benefice passed to his son, also Daniel, Beaufort's father. By the time Francis Beaufort was born in Navan, in 1774, the Beauforts were well established among the social elite of County Meath and Dublin. Daniel Junior was a talented, sociable and engaging man, with many talents, including that of map-maker. He published *The Grand Topography of Ireland*, one of the earliest detailed maps of Ireland, in 1792. This was widely popular but Daniel was a poor manager of money and constantly in debt. This prevented him from realising his potential as he felt he had to be continually on the move in order to stay just one step ahead of his creditors.

All the moving naturally had a destabilising impact on his family. They Beaufort children were dragged hither and thither in a constant attempt to save money and avoid creditors, so they were never in one place long enough to enjoy a lengthy period in any school and they were mostly self taught. When Beaufort was just two, the family moved from Navan to Mountrath, County Laois. They stayed there for three years but then – most probably with creditors again closing in – Daniel took the family across the Irish Sea to Wales, where they lived cheaply for a time near Carmarthan. Then came Chepstow in England, where Beaufort and his brother William were refused schooling on account of their 'Hibernian accents'. In 1784, Daniel's wealthy brother-in-law, Robin Waller, came to the rescue, organising funds so that the Beauforts could come back to live in Ireland. This time their destination was Mecklenburg Street in Dublin's north city centre.

In those years, just before the Act of Union of 1800 brought direct rule from London, many wealthy peers and members of the Irish House of Parliament had homes in Dublin. The north city centre, including the area around Mecklenburg Street (now Railway Street), was fashionable in the 1780s but gradually decayed after the Act of Union. In the 1780s, Mecklenburg Street was heaven to a sociable, educated man like Daniel Beaufort, who liked to entertain his peers in his fine home. Young Beaufort adored his father and inherited his wide range of interests and intellectual ability, if not his outgoing personality, and the conversation of the Dublin intellectuals who visited his home stimulated these interests.

There had been no navy men in the Beaufort family up to then and it seems that Beaufort ended up in the navy as a result of a chance event. His father had officiated at the marriage of a naval captain, who promised Daniel to give his son, Beaufort, some 'time' on his ship. This meant that the captain would be prepared to say that Beaufort had spent time at sea (although he hadn't) in order to give him a better chance of gaining entry to the navy and a career at sea – a common practice at the time. From that moment on Beaufort became fixated on a career in the navy.

It was during his time living in Dublin that Beaufort received his three years of formal education in a school long since defunct called the Master Bates Military and Marine Academy. At the academy, he studied hard and showed a natural ability for mathematics and draughtsmanship.

When he was fifteen, the naval captain came through on his promise and Beaufort got the opportunity to go to sea on board a ship called *The Vansittart*. He left from Gravesend, near London, on his first sea voyage. Beaufort impressed the captain of his ship, a man in his thirties called Lestock Wilson, who was a talented marine surveyor. He learned a lot about surveying from Wilson and Wilson's daughter, Alicia, was later to become his wife. In the early part of his career, Beaufort was keen to see combat, as he believed this would be the best way to climb the naval career ladder. He was also hopeful of bounty money, the fruits of plunder that were divided among seamen when an enemy ship was captured. In these years, his primary motivations were promotion and riches.

His first step up was to midshipman but his career advanced more slowly than he would have liked and in 1790, at the age of twenty-six, to his frustration, he was still a midshipman. His role in the 1794 'Glorious 1 June' naval encounter with the French – in which he fought bravely and got a 'ball' in his chest for his troubles that he carried with him for the rest of his life – helped gain him recognition, as did his exceptional talent for all things technical.

Beaufort's ship did not finally come in, as it were, until 1805, when he was thirty-two. In that year he was promoted to captain the *HMS Woolwich*. After he gained the captaincy he craved, his priorities shifted, or perhaps he no longer felt it necessary to attract the attention of his superiors. He showed himself more interested in peaceful pursuits, such as recording scientific data, than in combat. He steadily built a reputation as a superb map-maker, especially following an impressive survey of the southern Turkish coastline and its ancient historical sites. He was keen to introduce more reliable methods generally for recording data, including better means of taking depth soundings.

After almost forty years of naval service Beaufort was appointed Admiralty Chief Hydrography on 12 May 1829, just a few weeks short of his fifty-fifth birthday. He was delighted to get the job he coveted but he knew that he faced a huge task. The Hydrographic Office, established by George III in 1795, had a small staff and its activities were largely ignored by the naval brass but Beaufort wanted to put it at the centre of the Admiralty's thinking and planning. He transformed it into a body famed for producing the most trustworthy naval maps anywhere in the world.

When Beaufort took office, Britain was on the cusp of its industrial revolution, which brought energy and development across the fields of industry, science and engineering. The age of steam had arrived in 1825 with the first commercial rail service between Stockton and Darlington in the north of England. Goods were mass-produced in factories, and railways could transport these goods quickly to towns all over Britain. The production capacity of British industry was such that domestic demand did not match the supply of goods, and new markets were required. This was why the Empire, stretching from Canada to China, was vital to British interests. It provided an outlet for British goods and was the means of Britain's vast accumulation of wealth and power through trade in the Victorian era and into the 20th century. As Britain was a small island with a huge and far-flung empire, it was vital for international trade that charts be available to detail the safest sea routes. Beaufort aimed to fill what he called the 'woeful blank spaces' on the world's marine map, something he achieved over the following decades.

In 1829, the task ahead of him was enormous. At that stage, even the waters even around the British Isles had not been properly surveyed or charted, the only exception being the English Channel. This had received special attention as it was vital to Britain's interests to know as much as possible about the narrow stretch of water that separated it from France, its most feared enemy. The life and death struggle played out during Britain's 'Napoleonic' wars with France ended with the defeat of Napoleon in Waterloo in 1815, so naval ships that might previously have been earmarked for combat could henceforth be used for peaceful research purposes.

From a scientific point of view, Beaufort now held a position of immense importance. He was prominent in many of the leading scientific societies of the day, such as the Royal Geographic Society, the Royal Society and the Royal Astronomical Society. At this time Britain was the world leader in many fields of science, so these societies were among the most influential in the world. The Admiralty was the main supporter of science – the universities supported only a limited amount of research – so if a scientist wanted to get funding for his pet project, he often had to canvass Beaufort. He played a clever game by getting members of the scientific societies to suggest that the Admiralty should support a project (although he often drew up the details of the proposal himself). In this way, the Admiralty was seen to be doing the bidding of learned bodies but behind the scenes it was Beaufort who was pulling the strings. His power was enhanced by the fact that the navy gave its scientific branch, which included the Hydrographic Office, substantial autonomy to decide which projects it should support.

Beaufort contributed greatly to Britain's achieving superpower status by the mid-19th century. During his twenty-six years at the helm of the Hydrographic Office, hundreds of new, accurate maps were created. He made all the final decisions about projects and personnel, although he had to respond to current political or economic demands. For example, in 1851, during the Australian Gold Rush, it became important to survey the approaches to the 'Gold Coast, due to the volume of ships that were heading there from Europe and North America. A few years later, during the Crimean War which began in 1855, it became crucial to survey the Black Sea around the Crimea.

From the time he entered the navy at the age of fourteen or fifteen, a British naval officer was trained to be a meticulous observer and recorder of all manner of information. Some captains were very accomplished surveyors in their own right, yet Beaufort exerted a high level of control over the officers and their work. After he had selected the survey location and decided on the captain and boat to do the work, he would spend several weeks compiling specific orders for the captain to follow. His requirements were based on reading through the remark books of captains who had previously been in the area and any other relevant sources he could get his hands on. He also decided which scientific instruments would be taken on board and even on the stationery used. The captains were left in no doubt about what was expected of them on a particular mission. Beaufort's style of leadership was fatherly and the men under his command responded well to it. This in turn meant naval survey work that was consistently excellent, in home waters and worldwide, on Beaufort's watch.

Beaufort was obsessed with accuracy and would sometimes reprimand a captain for what he regarded as a lapse in surveying, recording or observing standards. One story suggests that the origins of this obsession can be traced to an occasion when he was shipwrecked as a boy sailor as a result of the ship's captain having faulty charts. Whatever the truth of this story, it was certainly true that, pre-Beaufort, the Hydrographic Office maps had a reputation for inaccuracy and omissions. Beaufort had both an artistic eye and mathematical logic, a powerful combination. By the time he retired, the British navy had mapped coastlines all over the world, taken better soundings of harbours (important for assessing the size of ships they could take), charted safer sea routes and identified the direction and force of dangerous winds. He also produced the first *Admiralty Tide Tables,* which were made available to all sailors, and pushed for the introduction of tide gauges around the English coast and beyond.

One of the key pieces of equipment on board a British naval surveying ship was its chronometer, a marine clock that could determine a ship's longitude. The chronometer was the equivalent of today's sat nav and an absolutely vital piece of equipment for any surveying boat. It worked by measuring the position of a celestial body, such as the moon, above the visible horizon, using a sextant. This gave the surveyor an exact time reading and, from this, it was possible to determine a ship's longitude and thereby its exact position by consulting published tables. It was vital for accuracy that chronometers were kept steady, regardless of ship movements or the sea's wildness, so they were mounted on suspension devices called gimbals. Beaufort constantly questioned captains as to whether the chronometers were working properly, as without them no accurate mapping could be done.

The other development that facilitated coastal map-making was the improvement in the technique of 'triangulation'. This makes use of a fixed point on land, such as a castle, which has a known location. The coastal survey ship moves up and down the coastline along a horizontal 'baseline' – a straight line – and never deviates from this path. The surveyor, at sea in his ship, charts the angle between his baseline location and the fixed point and draws a triangle, which will have two known angles and one known length. He can work out the other angles and lengths on the basis of this information, using some standard mathematical rules. A great number of triangles built up in this way and cross-referenced can be used to reduce the potential for surveying errors.

This was an exciting era, with Europeans discovering new lands. Captains took the opportunity to name islands or remote areas they discovered after prominent politicians or other influential people who might help advance their careers. This is why so many places are named after 19th-century British prime ministers. Beaufort did not like this practice, preferring names to be based on what native people called the place or on its physical characteristics. However, he was powerless to stop captains from choosing any name they pleased and he himself was often honoured by having his name attached to some place or other. On one occasion he changed a name from Beaufort Land to Queen's Land, as he felt it was impertinent for his name to be attached to the place. Nevertheless, there are places that still bear his name, such as Beaufort Island – a lump of volcanic basalt rock in the Antarctic Ocean – and the Beaufort Sea in the Arctic.

Captains were encouraged to bring scientists with them on their survey missions. Some famous names travelled with the navy to make recordings, such as Thomas Huxley, the zoologist, and William Hooker, later the director of Kew Botanical Gardens in London. However, the most famous by far was Charles Darwin, who made an epic trip to South America with Captain Robert Fitzroy on *The Beagle.* Fitzroy had suggested to Beaufort that it would be a good idea to bring a naturalist along on his next trip to South America as he had already done some surveys of Patagonia. It was a tremendous opportunity for a naturalist but the catch was that he or she would not be paid and had to travel under their own steam and resources. Beaufort put out feelers for Fitzroy. His friend and fellow-astronomer Thomas Peacock got in touch with a geologist called John Henslow and Henslow, in turn, suggested Charles Darwin. Darwin's observations of the plants and animals of the Galapagos Islands led to the publication of his revolutionary *On the Origin of Species by Means of Natural Selection* in 1859.

Beaufort and his wife Alicia (Wilson) had a very strong bond, despite the fact that Alicia did not share his scientific interests. After twenty years of marriage, Alicia got breast cancer. She died a few years later, leaving Beaufort inconsolable, with sole guardianship of his six children – two older boys, one younger boy and three girls – who ranged in age from twenty down to eight.. He asked his two spinster sisters, Harriet and Louisa, to come over from Dublin, where they lived together, to help him with his family in London.

Louisa was something of an intellectual and had written a book called *The Round Towers of Ireland.* Like her brother, she was a member of the Royal Irish Academy. But Harriet was

Beaufort's favourite. At the age of sixty-four, in 1838, he met and married Honora Edgeworth of Edgeworthstown, County Longford, who was twenty years his junior, and the stepdaughter of his sister Frances. His two sisters returned to Dublin and Honora embraced the London social scene with gusto, delighting in balls and parties. Beaufort had no interest in any of this and is quoted by author Nicholas Courtney in his book *Gale Force 10* as writing in a letter: 'How horrid dining abroad every day, wasting time, learning nothing, seeing less, eating and drinking too much, talking nonsense.' Honora had a stroke not long after she married Beaufort and another seven years later, after which she broke her hip and got an infection. Her health declined and she became disabled.

By 1855, the date of the outbreak of the Crimean War, Beaufort had become indispensable to the Admiralty. He was by now quite deaf and feeble and wanted to retire but the navy convinced him to stay on to help the war effort. He oversaw the drawing-up of strategically important maps for the Black Sea region, where that war took place. After the war he was allowed to retire but he retained his urge to observe and record on a daily basis and he kept a life-long weather journal.

Beaufort died in Brighton on 17 December 1857 and was buried in Hackney, east London, beside his first wife, Alicia. The great American hydrographer, Matthew Fontaine Maury, commented after his death: 'navigators of all nations owe him a deeper debt of gratitude than to any man dead or living.'

The Legacy of Francis Beaufort

- He was the greatest marine map-maker in history.
- He popularised the 'Beaufort Wind Scale'; a more accurate way to measure wind.
- His superb ocean charts helped making seafaring safer for all mankind.

- He was the first to chart accurately the world's oceans, seas, harbours and coastlines.
- He was a key figure in the rise of Britain to superpower status in the 19th century.
- He helped advance the fields of biology, meteorology, geology and astronomy.

2

THE SHAKING EARTH:

Robert Mallett (1810-81)

Born: Capel Street, Dublin

It's the autumn of 1849 and a terrible time for Ireland as it faces into its fifth winter since what became known as the Great Famine began in 1845. All over Ireland, people are sick, dying, or – the lucky ones – emigrating. Six miles south of Dublin on Killiney beach, thirty-nine-year-old Robert Mallett, a well-to-do engineer, businessman and scientist, has other things on his mind. He is on the beach with his son John, a promising seventeen-year-old chemistry student in TCD and about ten members of the Royal Irish Constabulary. The plan for the day is to blow up Killiney beach, using twenty-five pounds of gunpowder. The reason? Robert wants to test his revolutionary idea of why earthquakes occur by measuring the arrival of a 'shock wave' from the explosion site at another location half a mile down the beach with a specially built instrument called a seismoscope.

Ever since the devastating Lisbon earthquake of 1755, which estimates now suggest killed around 50,000 people and destroyed 85 per cent of the city, scientists all over the world have been trying to understand why earthquakes happen. It is one of the great scientific questions of the Victorian age and part of a general trend in science to try to understand the great forces of nature.

Ireland in 1849 is part of the British Empire and party to rapid industrial developments in railways, steam, iron, metal and coal. Science and technology are thriving and the United Kingdom of Great Britain and Ireland is the world's most technologically advanced nation. Despite these advances in knowledge, some strange notions of why earthquakes occur are widely held. In the 1840s the most popular scientific explanation is that earthquakes are somehow the result of coal burning inside the earth, or the result of pieces of the earth cooling and then cracking. There have been some interesting observations that could have led somewhere, such as the philosopher Kant noting that earthquakes didn't occur in flat countries, like his Prussian homeland, but in mountainous countries.

But no one grasped the significance of his observation. The reason was that there was no understanding of what the interior of the earth was made up of, or what was going on beneath the ground. This was a crippling intellectual weakness as the riddle of what caused earthquakes could be solved only by understanding what lies beneath. The inability of science to explain why earthquakes occur perhaps helps to explain why many people in the mid-19th century still believed that they were God's way of punishing the behaviour of sinners. This was the context in which Mallett was operating as he prepared his Killiney experiment.

Mallett's insight that earthquakes were caused by shock waves of energy travelling in an explosive fashion through rocks from a central point, or epicentre, where the energy originates, most likely came to him after conversations with William Rowan Hamilton, a fellow Dubliner and a superb mathematician. Hamilton suggested to Mallett that he apply the mathematical laws governing light to rocks. He reminded Mallett that light is diffracted, or bent, in different ways as it passes through different materials. Could the same thing be happening in rocks? Could the energy that causes earthquakes be travelling through different kinds of rock at different speeds? It was an intriguing possibility and Mallett began to think about how he could test it out.

In 1849 Killiney beach was quite remote from the city of Dublin so it was not too difficult to secure permission for an experiment in which twenty-five pounds of gunpowder would be detonated, leaving a huge hole in the beach – something that residents of today would hardly tolerate. The explosion was to take place at one end of the beach. This would replicate an earthquake and precisely half a mile away Mallett would be positioned with a specially-made piece of equipment to measure the arrival of its potential shock waves. The seismoscope Mallett built consisted of two viewing cross hairs that were fixated on a small dish of mercury liquid. The proof that a shock wave had arrived would be a ripple in the disk of mercury shortly after the explosion, like the effect of a small pebble hitting a still pond. It was a simple, ingenious experiment. If it worked, it would change forever how scientists understood earthquakes and other violent geological processes. It would also provide a clue to answering some questions that baffled Victorian scientists. Why did seashell fossils appear on mountain tops far from the sea? Why did damaging earthquakes repeatedly occur in specific areas of the world?

The fuse was lit, the sound of a massive explosion ripped through the cold air, a huge hole was blown in the beach and sand sent flying. Mallett heard the bang but didn't see it, as he was too busy peering down his seismoscope half a mile away. Then he saw the tell-tale circular ripples forming in the dish of mercury that signified the arrival of the shock waves. He was delighted. The theory of something he called 'seismic waves' travelling through rock and creating an earthquake had been proved correct. He had given birth to a new science, which he called seismology, the study of earthquakes. In the process he had shocked scientists into thinking about the earth in a totally new light, as a dynamic, changing entity.

Mallett was born in Dublin and grew up on the premises of his family's iron foundry, in Ryder's Row, off Capel Street. The Malletts were wealthy and mixed with the upper echelons of Dublin society. Mallett showed a great interest in and talent for science, particularly chemistry, from an early age. He spent long hours in his own personal chemistry lab as a child, using chemical mixtures and apparatus of various kinds borrowed from his father's ironworks. So much did he love being in his laboratory, performing experiments, that his parents would lock him out of his lab as a punishment for any childish misdeeds.

Mallett entered TCD at the age of sixteen, where he met like-minded souls such as William Rowan Hamilton. It appears that, although very clever, he wasn't the most diligent student and on one occasion he came close to being sent down. But he received his degree in natural science at the age of twenty and returned to work for his father.

The Malletts were similar to other well-off Protestant families in Dublin at the time. Their world was limited to contact with others of their kind and they had little or no knowledge of what was happening in the rest of the country at this time of hardship and famine. It is illuminating to note that Mallett reportedly reacted angrily when he was put out of an office that he had held near the Four Courts in Dublin, to make way for a famine relief agency. He belonged to a small, group that had the time and resources to pursue its scientific interests and curiosity.

The family firm, J. and R. Mallett, was one of the most renowned engineering enterprises in Ireland from the 1830s to the 1850s. It specialised in the manufacture of iron gates and railings – such as the iron railings that border the TCD campus along the length of Nassau Street, upon which the name Mallett can still be seen inscribed – but the company also worked on bridges, churches and even lighthouses.

Mallett lived with his parents through his college years but in 1836, aged twenty-six and by now well established as an engineer with the family firm, he moved to Glasnevin, on the north side of the city. He lived with his wife Cordelia in a house called Delville (where Jonathon Swift reportedly wrote *Gulliver's Travels* while staying with friends). The house was pulled down in 1944 to make way for a road. The Malletts had six children, three boys and three girls, but tragically, in 1854, Cordelia died. A few years later Mallett moved to Kingstown (now Dún Laoghaire), where he took up residence in 1 Grosvenor Terrace.

Mallett had such excellent engineering skills that he could turn his hand to almost anything. His projects included a swing bridge over the River Shannon at Athlone which allowed for river crossing and the passage of boats; the construction of the lighthouse at Fastnet Rock eight miles off the coast of Cork; and the lifting and reinforcement of the roof of St George's Church in Hardwicke Street, Dublin. When Mallett began the work to restore the roof, the walls of St George's were bulging. The original roof beams had been too short for purpose and the whole structure was close to collapse. He persuaded the church authorities that he could lift the roof – in situ – insert iron beam arches to support it and drop it back into place. The two-hundred-foot-tall spire of St George's is still a landmark in Dublin's north inner city. Mallett's reputation as an engineering genius grew from this. If there was a difficult or tricky engineering job to be done, he was the man.

After his Killiney experiment Mallett began looking for an opportunity to study the impact of an actual earthquake close up to test his seismic theories in the field. This opportunity arrived in December 1857 when a huge earthquake, the third biggest on record, hit Italy, close to Naples. Many towns and villages in the area were destroyed. Official figures estimated 12,000 dead but some unofficial sources suggested that closer to 200,000 people had died as a result.

When news of the Naples earthquake reached Mallett he applied to the Royal Society, an esteemed scientific organisation based in London, for a travel grant to go to Italy and study the earthquake. Mallett had presented his theories of why earthquakes happen to the Society on several occasions and his successful experiment in Killiney had enhanced his standing. He was by this time aged forty-seven. A wealthy man, he was living a life more akin to that of a gentleman scientist than that of a hard-working businessman and engineer but he applied to the Royal Society for the equivalent of about €5500 in today's money to fund his trip to Italy. After some wrangling, the Society agreed to support his proposal for a scientific study. Mallett had the backing of two giants of science, Charles Darwin and Sir Charles Lyell, the famous geologist who popularised the notion that the earth had been shaped by slow-moving forces that were still at work.

Equipped with his travel grant, Mallett made his way to southern Italy, getting there about two weeks after the earthquake. He arrived in Naples, bought three mules and acquired the services of a French photographer called Alphonse Bernoud, an inspired move. The science of photography was still in its infancy in 1857 but Mallett saw its potential to record precisely the aftermath of an earthquake. In his grant application he included the cost of hiring a photographer and publishing the photographs as part of his final scientific report. The Society baulked at supporting the cost of photography, as photographs were expensive to develop and publish and at this time scientific reports didn't normally include them. But Mallett stood firm on this issue and the result was the first photographs ever taken of the aftermath of an earthquake. It was a visionary move, breaking new ground for scientists.

Not only were Mallett's the first photographs of an earthquake, they were produced in 3D, something extraordinary for the time. The photographer, Bernoud, was asked to shoot his images stereographically; that is, to shoot two photos of the same scene. Mallett then used a stereoscope to blend the two images into one three-dimensional image, which provided more intricate detail of the devastation.

The two men stayed in a village called Pertosa and from there they made daily trips to record the damage caused by the earthquake. The Neapolitan earthquake had a magnitude of 8-8.5 on the Richter Scale. It was very shallow, meaning that its epicentre was close to the surface, causing a lot of ground shaking. Buildings that had been constructed without cement collapsed into a pile of rubble, burying thousands of people. At this time there was no Red Cross, no emergency field hospitals or international relief teams sifting through the rubble for bodies.

Mallett must have been witness to awful scenes of death and suffering everywhere he went but his published report studiously ignored human tragedy, focusing exclusively on science. He took copious notes as he moved around the countryside, noting, for example, how religious statues had been twisted and turned in relation to their pre-earthquake orientation. The over-arching aim was to look closely at the damage in order to determine the direction from which the major shock wave hit the area, then, from many such observations, to try to determine the epicentre. This was the first time such a scientific approach had been taken to studying the impact of an earthquake and it's the approach still used today.

The precision of Mallett's notes and drawings of buildings was remarkable and showed his superb abilities as a civil engineer. For example, he stayed overnight in the beautiful monastery of Certosa di San Lorenzo in a town called Padula, which today is a UNESCO World Heritage Site. In 1857, the monastery was right in the path of the deadly earthquake but it was the least destroyed building in its area. That was why Mallett decided to stay there. He produced a perfect architectural drawing of the church, overlaid by the cracks resulting from the earthquake. This level of precision helped him to unravel the earthquake's origins and direction.

It took a few years for Mallett to gather a vast amount of data, including notes, photographs and drawings, before he could present his findings in a coherent fashion to his patron, the Royal Society in London. His presentation, in 1862, had the title 'Great Neapolitan earthquake of 1857: The First Principles of Observational Seismology'. This was the official birth of seismology – the science of studying earthquakes – a term Mallett coined. It swept away all the crazy theories of why earthquakes happen, both scientific and religious. It was a highly numerate report, with numbers quoted down to eight places of decimals, extremely precise and reflecting Mallett's skill as a civil engineer.

Mallett discovered what caused earthquakes but he also came tantalisingly close to solving the bigger puzzle of why earthquakes occur far more regularly in certain parts of the globe than in others. This question was not answered until the 1960s, with the theory of plate tectonics.

As a follow-up to his Killiney experiment he had decided to map precisely the distribution and intensity of the world's earthquakes and put all the information on what's called a Mercator map. This was a type of map projection that had been developed by the Belgian cartographer Gerardus Mercator in 1569, which served to increase artificially the size of Europe and therefore its global importance. The Mercator approach provides the view of the world that most Europeans still have today.

The Mercator map that Mallett drew up was fascinating. Anyone who has studied geology at university, or perhaps even at secondary school, will immediately notice something when they look at it. The lines linking areas of equal earthquake activity and intensity on Mallett's map – which he called seismic lines – bear a striking resemblance to the boundaries between the earth's tectonic plates, the pieces of crust that collide to create earthquakes. In this way, Mallett came close to understanding plate tectonics, the theory of which revolutionised geology in the 20th century, proposing that continents were sliding past, over and under one another.

According to plate tectonics the earth can be thought of as being like an egg that has been boiled but is still soft inside. Think then of the surface of the egg, the shell, being dropped and cracked. The cracked pieces resemble the tectonic plates, riding on a molten interior. However, unlike the egg, the plates of the earth are in constant (albeit very slow) movement, grinding past one another, sliding under or over one another or slamming headlong into one another, creating earthquakes, building mountains and melting and pressuring rock to form volcanic magma.

Mallett lacked crucial data that would have completed the puzzle. He had information on earthquakes from reports all over the world, even from the Azores in the Atlantic Ocean, but he had no means of recording earthquakes that occurred at sea. As any geologist will tell you, one of the most important plate boundaries in the world is the mid-Atlantic ridge, which runs through Iceland and down the middle of the Atlantic. Mallett would have been completely unaware that such an undersea ridge existed.

If he had had this information, he might have realised that there were pieces that fit together and that earthquakes occurred along the boundaries. He might also have noticed that South America could be fitted nicely into Africa, like two pieces of a giant puzzle. This might in turn have led him to conclude that the continents were all moving. But the solution to the riddle would have to wait another century or so.

As to where Mallett's original Mercator map is today – unfortunately, no one knows. Mallett bequeathed it to the Royal Society but it appears that they subsequently lost it. Tom Blake of the Geophysics Unit in the Dublin Institute of Advanced Studies is determined to bring home the map as part of Ireland's scientific heritage but he hasn't managed to locate it yet.

By 1860, engineering work had slowed down in Ireland. The main railway routes had almost been completed and there was huge competition in ironworks from lower-priced, more specialised iron foundries that were springing up all over England and Scotland. R. and J. Mallett could no longer compete, as the cost of importing raw materials, iron and coal had put them at a disadvantage compared to engineering firms on 'the mainland'. Irish industry was too small to provide enough business for the firm to survive.

The crunch came when an expected contract from Dublin Corporation to make water pipes and other bits and pieces associated with new waterworks went to a Scottish firm. The decision was made to wind up the family business.

After the firm closed, Mallett decided to move to London, where he was now spending a lot of his time anyway, doing research and attending meetings. He married again in 1861, at the age of fifty-one, and took up residence in a fine house on Clapham Road to the south of the city. In London he worked as an engineering consultant and was active in the Institute of Civil Engineers. He was something of a visionary. For example, as president of the institute, he is on record as stating that Ireland should import the technology then available in Britain, develop its own industry and become less dependent on agriculture.

In the winter of 1871 and 1872, when he was in his early sixties, Mallett's eyesight started to deteriorate. Working on his own became virtually impossible and he had to employ someone to whom he could dictate his notes. He was ill and confined to his bed for a year before his death, on 6 November 1881.

The Legacy of Robert Mallett

- He was the founding father of seismology, the study of earthquakes, and is credited with inventing the term.
- He discovered that earthquakes are due to shock waves passing through rocks.
- He performed the first scientific study of the aftermath of an earthquake following the Great Neapolitan Earthquake of 1857.
- He was the first to use photography as a tool for recording earthquake damage.
- He laid the foundation for what became the theory of plate tectonics.
- He helped establish earth science on a rational and scientific footing.

3

ELECTRICITY FOR ALL:
Nicholas Callan (1799-1864)

Born: Darver, Dundalk, County Louth

It is hard to imagine a world without electricity; desk-top computers, mobile phones, televisions, cookers, washing machines, home heating systems and fridge freezers. Cheap, abundant electricity is one of the pillars on which our modern world is built. We are all – in the developed world at least – utterly dependent on it and life without it – even fleetingly – feels like an unwelcome descent into the Dark Ages. We have all felt the irritation when the electrical power goes off without warning, the TV goes blank, darkness descends and a frantic search for candles and matches begins.

The credit for the miracle of producing 'electricity for all' goes to a small number of gifted 19th-century electrical scientists. The list includes famous names such as Michael Faraday and Thomas Edison, along with an obscure Irish priest and professor at St Patrick's College, Maynooth. This was Nicholas Callan, born into a well-to-do farming family in County Louth. Callan was an experimental genius who has not received the recognition he deserves for his breakthroughs in electrical science, notably the invention of what was called the induction coil. This device was the world's first electrical transformer, opening the door to development of modern grid systems, in which voltages must be greatly increased during the transport of electricity, via wires, from the point of power generation and then greatly reduced before reaching users in homes, offices and workplaces. This is what a transformer does: it steps up or steps down voltage as needed.

As a scientist based in a primarily ecclesiastical college, Callan worked in virtually anonymity. Many of his clerical colleagues in Maynooth had little knowledge of his work, while some of those who did told him he was wasting his time. However, Callan always kept faith in his work. He had a clear vision of how his coil could help bring electricity to the masses, even in the early days after its invention in 1837. Many scientists of this period were not even aware that such a coil had been invented or that it removed a technological roadblock, by making it possible to ramp up voltages and ensure that electricity could be distributed over long distances with little loss of energy en route.

It wasn't until the 1950 publication of *Nicholas Callan, Priest-Scientist (1799-1864)* by P.J. McLaughlin that Callan began to get some credit for his work in the field of electricity. Fr McLaughlin was appointed to Callan's old post of Professor of Natural Philosophy at Maynooth in 1928 and his book provided an overview of Callan's scientific strategy, his breakthroughs and his place in the bigger picture.

'He focused on two aspects of electricity, what he called quantity electricity and what he called intensity electricity,' wrote McLaughlin. 'He aimed to produce at low cost large amounts of both sorts. He aimed also to find out how to convert one into the other. In doing so, he discovered the principle of the step-up transformer, the first rung on the ladder to high-tension electricity and to electricity as a public utility among large communities.'

We have known about electricity, this invisible power source, for a long time, at least as far back as 2750BC, when ancient Egyptian writers described how people were 'shocked' when they touched some of the fish taken from the Nile. Around 600BC, Thales of Miletos, a Greek philosopher, observed what we know today as 'static electricity' and – wrongly as it turned out – theorised that it was caused by magnetic effects. But it was not until the 19th century that researchers began to understand what exactly electricity was and how it could be manipulated.

The 19th century – the 'electric century' – was the era in which most of the important electrical discoveries were made. The century began with Volta's battery in 1800, followed in 1826 by Ohm's Law, in which George Ohm (a Bavarian physicist) defined the relationship between power, current and resistance. This law said that the electrical current passing through an electrical conductor between two particular points will be dependent on the level of resistance to the current that exists there. This was the first time electrical resistance had been defined in mathematical terms.

In 1831, Englishman Michael Faraday showed that electricity could be produced by magnets and created an electromagnetic field. The electrical relay invented by American Joseph Henry arrived in 1835: the first device that could send electrical current over long distances. Samuel Morse's electric telegraph, invented in 1844, could send messages long distances using wires, via what later became known as Morse Code. In 1880, Thomas Edison, the great American inventor, produced the world's first long-lasting light bulb, which could illuminate for up to 1200 hours without failing. In 1884, Irishman Charles Parsons also contributed to the story of electricity when he invented a steam turbine that could generate large amounts of electricity. The century closed with the opening of power plants, such as the Niagara Falls Hydropower Station (1896), which supplied electricity to the city of Buffalo in New York State. Inside just one century, mankind had mastered the production and use of electricity.

Where does Callan fit into this picture? Well, he plays a central part in it, primarily through the invention of the induction coil. Even today, electrical transformers are essential for the widespread distribution of electricity over very long distances. The idea underpinning any electrical grid system is to transport the electricity at high voltages and make it available at low voltages. The laws of physics mean that the higher the electrical voltage during its transport, the less energy is lost. The voltages must be increased where electricity is generated before it enters the grid.

In Ireland, electricity is transported along about 6500 km of high-tension wires at between 110,000 and 400,000 volts. These lethal voltage levels are reduced enormously to between 220V and 240V before electricity enters the home. This grid system requires the deployment of step-up and step-down transformers to control how electricity is distributed. Step-down transformers are contained in the boxes that can be seen perched up on top of electricity poles. Callan's coil was a superb step-up transformer. He took the voltage in a 6V battery and ramped it up to 600,000V. The scale of the technical achievement of this brilliant but modest priest can be assessed by pointing out that almost a full century later, in the early 1930s, well-funded research teams in the US and the UK were struggling to reach similar high voltages as part of the race to 'split the atom'.

Callan was born in 1799 in a small village called Darver, just south of Dundalk, into a prosperous family of big farmers, who also dabbled in baking, brewing and distilling. This comfortable background would prove important to Callan's scientific success in later life, as he funded his research from his private means after he became Professor of Natural Philosophy (Physics) at Maynooth University in 1826, at the age of twenty-seven. Callan was at his most creative and productive in the 1830s and 1840s when there was very little money available for things like lab equipment and chemicals. By funding his research from his private means, he was able to hand over his entire salary to help those who were suffering from hunger and disease.

Somewhat surprisingly for someone who became a Catholic priest, Callan was educated, early on, by William Nelson, a Presbyterian clergyman in Dundalk with a good reputation as a teacher. Callan then attended a seminary in Navan before entering Maynooth in 1816, aged seventeen. He spent the next forty-eight years there as student and teacher.

In Callan's time and for many decades more, Maynooth was thought of as an ecclesiastical college first and foremost. This view was not entirely fair, as science had been taught at Maynooth since the college's foundation in 1795. The British government set up Maynooth because they were concerned that Catholic priests might pick up revolutionary ideas while receiving an education in Irish colleges in locations across the Continent, such as Paris and Salamanca. Seven professors were brought in from the Sorbonne in Paris to establish Maynooth and one of these was a Professor Darley, who became the university's Professor of Natural Philosophy. By the time young Callan came along, Cornelius Denvir was professor. Denvir had a huge influence on the scientific development of the young seminarian. He schooled him in the new experimental method of science, which sought to demonstrate and prove theories in the laboratory, and introduced him to electricity and magnetism, two areas that consumed much of his later life.

Callan was ordained in 1823, then went to Sapienza University in Rome where he was awarded a PhD in Divinity. During this time he became familiar with the work of Volta and Luigi Galvani, who discovered that bioelectricity resides inside animals, while doing experiments on frogs. As Callan finished his PhD, opportunity knocked. Denvir was appointed Bishop of Down and Connor and Callan replaced his mentor as Professor of Natural Philosophy in Maynooth.

Some surviving anecdotes hint at Callan's wicked sense of humour. One story concerns a visit he made to view the (then) world's largest telescope which had been completed by the Third Earl of Rosse in Parsonstown in 1845 (*see* Chapter 12). For reasons unknown the Maynooth priest was refused admittance to see the telescope and went home disappointed. Later, Lord Rosse reportedly came to Maynooth and asked to see Callan's famous induction coil. The story goes that Callan refused this request and suggested that Rosse might return home and view the coil through his telescope.

Then there are stories about seminarians who were used as guinea pigs to test the voltage levels during some of Callan's rather eccentric electrical experiments. At the time, no instrument could test voltage levels and Callan wanted to test how much voltage was present after it had been boosted in his induction coil. Famously, he got fifteen clerical students to hold hands, with the last two having the thankless job of putting their hands on the electrical output of the induction coil. Callan assessed the voltage by watching how high these last two jumped.

Things got out of hand when one seminarian, a future Archbishop of Dublin called William Walsh, was knocked unconscious by a strong shock. After that, the authorities forbade Callan to experiment on students so he adapted his experiments and used turkeys instead. He attached electrodes to a turkey's head and if it was killed by electrocution, the voltage was deemed to be good.

Callan was indefatigable in his determination to find ways to generate large electrical currents and high voltages. To this end he built huge batteries and electromagnets. An electromagnet results when an electrical field is produced around a central piece of metal. The amount of electricity produced can be determined by assessing the strength of the magnet. One of Callan's electromagnets was capable of lifting a weight of twenty tonnes, the equivalent of eight white rhinoceros, fourteen hippopotami or forty gorillas.

One of the keys to Callan being able to generate such large voltages in his coil and building very powerful electromagnets – the most powerful in the world at the time – was his invention of the first mechanical current breaker. An American, Joseph Henry, had built a 'shock machine' by interrupting electrical current. He discovered that breaking the current and re-establishing it served to build up the voltage. Henry's machine was crude and its current was broken by hand, using an unfortunate assistant who got shocked as result. Then Professor McGauley of Dublin introduced an automatic hammer break system for the same purpose, like the type used in an ordinary electric bell.

Callan made significant improvements to McGauley's device. He developed a 'repeater', which was a far more reliable way of breaking and controlling an electrical circuit. This made use of the mechanism of an old grandfather clock to interrupt the current very rapidly, using a combination of copper wires dipping into cups of mercury. The current was broken every time the wires went into the liquid. With this apparatus Callan showed that more rapid current breaks increased current, with the repeater creating an intense current of electricity. He demonstrated how much electricity he could generate with his repeater by attaching the electrodes to the head of a turkey and electrocuting it. No one had ever come near to producing electricity of this lethal strength.

It could be said that two men, either Irish or of Irish extraction were critical for the birth of the mass-production car industry: Irish-American Henry Ford, the business genius behind Ford Motor Company, and Callan. Up to 1926, all motor cars had to be cranked up by hand in order to get them started. There were no batteries in cars and no automatic starting mechanisms. This made it impossible for anyone without a substantial amount of physical strength to consider owning a car.

The first car to have a battery and an automatic starting mechanism, operated simply by turning an ignition key, was the 1926 Model T Ford. The technology that made the starting mechanism possible was based on Callan's induction coil, which had been invented eighty-nine years earlier. The coil provided car engineers with a way to massively ramp up the power supplied by the Model T's small battery. The added power meant that sparks could be created that ignited the petrol in the car and sparked the pistons: this in turn drove the crank shaft and powered the engine into life. The development of cars with a simple turn-key starting mechanism made it far easier for people to operate them and as a result they became more appealing to a mass market.

P.J. McLaughlin did more than anyone to reintroduce Callan to the world of science in the 20th century. He described him thus: 'In character, Callan was an amiable man, modest and retiring, grave and absent – yet not without genuine humour and the power, when needful, of self-assertion. He was always a worker and when he died, he left some important acquisitions behind him to add to the scientific wealth of the world.'

A passionate experimentalist, Callan always gave the impression when he was away from his lab that he was impatient to get back to it. He was not very interested in communicating the results of his research to the outside world, preferring to get on with his work unlocking the secrets of nature. He would tell friends about his discoveries but neglect to inform scientific publications, as his mind had already returned to further laboratory work.

This meant that Callan received no recognition for his work until late in life, when he finally gave in to his friends' urgings to publish the results of his research and establish credit for and ownership of his work. It also left him vulnerable to other people wrongly claiming credit for inventions that belonged to him. The best example of this was the induction coil itself. After Callan had invented the coil in 1837 he essentially sat on the discovery and didn't report it for years, decades in fact. The discovery slowly became known to specialists such as an instrument maker in Paris, Heinrich Ruhmkorff, who began making induction coils and marking the instruments with his name. This was standard practice for instrument makers and in this case it resulted in people calling Callan's induction coil 'the Ruhmkorff coil' a name that stuck.

Finally, in 1857, twenty years after he invented the coil, Callan decided to set the record straight, by writing a paper for a publication called *The Philosophical Magazine* in which he claimed ownership of the induction coil. In the long and interesting paper Callan explained, among other things, the coil's potential to improve the operation of the Atlantic telegraph, which required high voltages for its long-distance transmissions. He also wrote intriguingly of the coil's potential 'for producing electric light'.

It is sad to think that after his death, Callan was largely forgotten, even in Maynooth, where many of his clerical contemporaries had shown little interest in his work. It is not surprising, then, that his equipment and apparatus were allowed to fall into decay after his death. They were rescued by McLaughlin, who set about cataloguing and systematically moving Callan's most important pieces of equipment to the museum in Maynooth, now called the National Science Museum.

McLaughlin was also keen to restore the credit for the induction coil to Callan. He went through the scientist's papers and proved beyond doubt that he had invented the coil in 1837, fifteen years before Ruhmkorff started producing his coils. The induction coil was finally credited to its true inventor.

The Legacy of Nicholas Callan

- He developed machines to generate alternating electricity current, a form of electricity crucial to the efficient distribution of electrical power.
- He invented the induction coil, the world's first step-up transformer.
- His work helped ensure that electrical power became available for all.
- He made significant improvements to battery technology.
- He patented a method of protecting iron from rust, an early form of galvanising.
- His technology was used in the first automatic car-starting mechanism.
- He built the first successful current breaker or circuit breaker.
- He constructed the most powerful batteries and magnets of his time.

4

THE LITTLE ICE AGE:

Annie Maunder (1868-1947)

Born: Strabane, County Tyrone

London, January 1709. The magnificent St Paul's Cathedral, some 365 feet tall, designed by the architectural genius, Sir Christopher Wren, has been rebuilt and reconsecrated after the Great Fire of 1666. If any visiting dignitaries peer down from the colossal dome of the new St Paul's towards the nearby Thames, they will not be at all surprised to see that, once again, as has happened repeatedly over recent decades, the river is frozen solid. Neither will they be surprised to see large groups of giddy, cheerful people milling about on the ice, along with hawkers selling all kinds of wares, as well as puppet shows, ice skating, parties, drinking and people being hauled across the ice in 'ice taxis' shaped like contemporary wooden ships.

This is a London 'frost fair' and it has become a major mid-winter social event in the city. However, this winter is even colder than Londoners have come to expect, with the night-time temperatures dropping to -12°C. What the locals don't realise is that they are living through England's and Europe's coldest winter in five hundred years, the coldest in a period that scientists would later call the 'Little Ice Age' (1645-1715).

Nearly two centuries later, two talented solar astronomers, Englishman Walter Maunder and his Tyrone-born wife, Annie, working together in the Royal Greenwich Observatory (also located on the bank of the Thames) found increasing evidence that the Little Ice Age coincided with an almost complete lack of sunspots (caused by intense magnetic storms) on the sun. The less stormy the sun was, it seemed, the less heat it gave out. This period became widely known as the 'Maunder Minimum', a name given to it in a landmark research paper by American astronomer Jack Eddy, which was published in 1976 in the US journal *Science*.

The 'Maunder Minimum' was Eddy's way of recognising the pioneering solar work of the Maunders. In the first half of the 19th century, German astronomer Gustav Sporer, who was interested in studying sunspots, had examined records of sunspot observations from reliable witnesses, going back to the early 1600s. Sporer noticed that the earth's cold periods coincided with a lack of sunspots. The Maunders were familiar with Sporer's work. They too looked at historical reports and added compelling data from thirty years of sunspot observations in Greenwich, both their own observations and those of their predecessors in the department.

The case for a sun-earth climate link as first outlined by Sporer and strengthened by the Maunders, was, therefore, well known to astronomers long before Eddy's famous paper appeared in *Science*. The term 'Maunder Minimum', was already in use by specialist solar scientists before 1976, yet it became widely used only after Eddy's paper. Eddy added some new evidence of his own to support the Maunders' central thesis that the sun's natural cycles had an impact on the cooling and warming of the earth. Eddy's claim was of great interest to many people in the 1970s, just as the environmental movement was starting to find its feet. Today, the interest in the subject is even more intense as it concerns an issue – global warming and its causes – central to the survival of mankind on earth. The solar investigations of Sporer, the Maunders and, more recently, Eddy, all remain at the cutting edge of the climate debate. The Maunders' work means that global warming, as seen since 1950 or so, cannot be entirely blamed on man-made greenhouse gases. The sun's cycles must also be taken into account.

The fact that Annie Maunder, née Russell, made it into such exalted scientific company in the Victorian era is hugely impressive given the obstacles that women wishing to become professional scientists faced at that time. She was born in Strabane, County Tyrone, in 1868, and was, by anyone's estimation, an extraordinary woman. She and her siblings were all bright and ambitious and their parents encouraged them to reach their potential but even in the context of this kind of family her career was exceptional.

This was an era when it was considered unseemly for women take part in scientific research on an amateur basis, let alone as professionals. Women had only recently been allowed to study science at university but regardless of the academic standards they achieved they could not be awarded degrees. Annie Russell gained admittance to a leading college, Girton College, Cambridge, excelled there and went on to a job that paid her (poorly) for what she loved doing: astronomy. She suffered further as a woman astronomer ahead of her time. She could not publish any research under her own name, was refused admittance to the professional body of astronomers, the Royal Society, and had to resign from work (officially at least) when she married her boss, Walter Maunder, in 1895.

Despite these hurdles Maunder left an impressive legacy. She broke through many of the barriers that stood in the way of talented female scientists who wished to become professional researchers. She was one of just two women astronomers working in the Royal Greenwich Observatory in 1891. She was also one of the first women elected as a Fellow of the Royal Society in 1916, twenty years after she was first proposed, when the ban on women Fellows was lifted.

Maunder's scientific legacy was equally important, as she and her husband were among the foremost solar scientists in history. She was better educated than Walter, so it is likely that she played a leading role in many of their co-publications. The Maunders were a powerful scientific team both before and after they married and their observations and analysis contributed greatly to what is known about the sun.

Astronomers had been looking at the sun and its sunspots since the early 17th century. Galileo Galilei is credited as being the first to use a telescope to study the stars in 1609 and there are good records of sunspot observations right through the 1600s. The reports get interesting from about 1645 onwards, when sunspots became harder and harder to find. Soon, they had almost disappeared and things stayed that way until 1715. The Maunders analysed the solar data from the period and put forward the idea that the cold was due to reduced magnetic storm activity on the sun, which meant that its heat output was reduced. This was a revolutionary idea as it suggested that the sun was a changeable star, whose output varied over time, challenging the consensus that the sun's output was stable.

Gustav Sporer was born in 1822, a generation before Walter Maunder, and he was the first to notice that sunspots varied over time. Walter Maunder was also interested in sunspots and when Annie Russell began to work at his solar department at Greenwich in 1891, after completing her education in Girton, she shared this interest. One of the most valuable pieces of work Sporer had done was to use mathematics to show how sunspots varied over time, in their solar latitude and size, depending on which part of the sun's cycle they appeared at. This was called 'Sporer's Law' and the Maunders' observations of sunspots, their size, location and occurrence, supported it.

The Maunders thus built significantly on the trailblazing work of Sporer and provided more observational data to support the sun-earth climate link. In his famous paper of 1976, 'The Maunder Minimum', Jack Eddy, as well as citing reliable sunspot data, decided to look at historical reports of the aurora borealis, or northern lights, from previous centuries. This was important because the spectacular light-show created by the northern lights – there are also southern lights, commonly called aurora australis – is strongly linked to sunspot activity. When sunspots are large, it means that magnetic storms are occurring on the sun and that charged particles are being spat off its surface and fired towards the earth. When these high-energy, charged particles reach the earth, they get caught up in its magnetic field, which is called the magnetosphere. The charged particles are sucked, like a piece of metal towards a magnet, towards the magnetic poles located at very low latitudes in the northern and southern hemispheres. It's not surprising that the charged particles are drawn towards these locations where magnetism is strongest. There, the interaction of the charged particles with the atmosphere causes the sparkling, colourful lights. In this way, historical reports of northern and southern lights also indicate sunspot activity.

Sunspots are visible to the naked eye at both sunrise and sunset and Eddy took into account sunspot records from Asia made without the use of telescopes. He took sunspot science to another level, beyond what was possible for the Maunders. For example, he measured the concentrations of Carbon-14 in the growth rings of trees. The presence of this kind of carbon in something indicates that the something – usually organic (previously living) – has been subject to a high level of solar activity. The amount of Carbon-14 present in lake sediments, for example, as well as in tree growth rings, provides a precise basis for assessing the level of past sunspot activity.

The annual rate of tree growth can be determined by cutting down a tree and looking closely at the width of each growth ring. Some rings are wider than others. Scientists have built tree ring records for every year going back thousands of years by comparing the tree rings from trees all over the world, living and dead. A poor growing season is reflected in a narrow tree ring in all trees alive in that year, while a wide ring indicates warm conditions conducive to good growth in all trees that were living in that year.

Carbon-14 is important because it is produced in the earth's atmosphere in response to periodic bombardment by the sun's cosmic rays. These rays are strong during magnetic storms, when the sunspot activity is highest. They cause nitrogen gas in the atmosphere to disintegrate, or break down, from Carbon-12, the form that makes up 99 per cent of all carbon, into Carbon-14, which is rare and occurs only in trace amounts. If a particular annual tree ring has a lot of Carbon-14, this indicates that there was a lot of solar activity in that year and lots of sunspots. If there is little Carbon-14 present, the opposite is true. Eddy found a scarcity of Carbon-14 in tree rings from the period of the Maunder Minimum, supporting the view that it was caused by solar factors.

Annie Russell was the eldest daughter of William Russell and his wife Hester, 'Hessy' (neé Dill). She had a religious upbringing, not surprising as her father was the Presbyterian minister in her hometown of Strabane and her mother was the daughter of a Presbyterian minister. There were six children in the Russell household, two sons from her father's first marriage and two sons and two daughters from the second.

Russell's parents sent her to the Ladies' Collegiate School in Belfast, one of the best regarded girls' schools in Britain or Ireland at the time (it was renamed Victoria College in 1887). The school was founded in 1859 by Margaret Byers, an advocate for women's education and a supporter of the women's suffrage movement, and very soon established a reputation for academic excellence. Pupils were taught subjects like current affairs and languages, which weren't available in most other girls' schools.

(This tradition of encouraging girls in all subject areas continues and in 2009 Victoria College was awarded 'specialist school' status, in recognition of its focus on the STEM subjects of science, technology, engineering and maths.)

At school, Russell's academic ability was clear and in 1886 she won a prize in the Intermediate Examination, a public examination which was then taken by students all over Ireland. She had the opportunity to prepare for the entrance exam for the Queen's universities in Cork, Belfast or Galway but she decided to take another route and sat – without preparation – the entrance exam for Girton College, a women-only college that was a constituent part of the University of Cambridge. Girton was a renowned place of learning for women but its students still suffered great inequality. They were permitted to sit the University of Cambridge degree exams but, even if they excelled, they would not be awarded degrees. Girton was not granted full and equal membership of the university until 1948.

Russell did well enough in the entrance exams to earn herself a £35 annual scholarship and three years later, in 1889, she came top of her class in mathematics, distinguishing herself by achieving the highest course placing for an Irish woman student in the University of Cambridge. Her mathematics tutor, W.H. Young, conveyed her ultra-determined personality in this slightly patronising reference to her: 'More than ordinarily handicapped – even for a woman – by an insufficiency of preliminary training, nothing but the power Miss Russell has of throwing herself completely into her work, could have enabled her to read as far as she has and with such success.'

After graduating, Russell spent a year teaching mathematics in a girls' school, the Ladies' High School, Jersey. She was quite shy and didn't like public speaking, which might explain why she soon came to believe that teaching didn't suit her. She heard that the Royal Observatory in Greenwich was employing women for the first time ever, albeit at a rank far below their abilities. One of her friends from Girton, Alice Everett, who also came from what is now Northern Ireland, was already working there.

Russell was offered a post in Greenwich Observatory but before accepting it she wrote to say that the salary of £4 per month was hardly enough to live on. She had been earning more as a teacher in Jersey. The observatory replied that it was offered on a 'take it or leave it' basis and if Russell's main interest was in the salary, perhaps she shouldn't accept the job. Despite the pitiful pay, which wasn't much more than her student grant at Girton, she began work there on 1 September 1891: at least the job was a unique opportunity for a woman trained in science and maths to work professionally in one of these areas. Up to then, research posts were not available to women (although there may have been a few rare exceptions) and this probably explains why she initially went into teaching. While women were treated unfairly, the fact that Greenwich took women astronomers on at all was significant and progressive for the time.

The rank at which Russell and her friend Alice were employed was that of 'lady computer', the same level as 'boy computers' who had started work there at fourteen or fifteen without university education. It was a frustrating situation for the two talented young women. The main duty of the 'computers' was to do the menial calculations that the more senior male astronomers – who got to observe the stars through telescopes – did not want to bother with.

There were various scientific departments in Greenwich and Russell was assigned to the solar department, to work under her future husband, Walter Maunder. Her daily routine involved taking detailed photographs of the sun and working out the solar coordinates and size of the sunspots. These images were supplemented by others taken in India and Mauritius. Together Maunder and Russell analysed the images and came up with their famous 'butterfly' diagram. This diagram, which looked like three butterflies drawn on a page by thousands of tiny dots of a pen, conclusively proved what Sporer had first suggested: that the appearance, disappearance and movement of sunspots were based on a repeating cycle. At early points in the cycle the sunspots would be at low latitudes, while at later points they would drift up to high latitudes. Then they would disappear and the cycle would start again as they reappeared at high latitudes.

The outstanding quality of Russell's photographs and her analytic skills were of huge help to her hard-working, observant boss. She photographed the sun every day, weather allowing, developed the photos and examined the negatives using a micrometer, a mechanical tool with a calibrating screw used by engineers to calculate very small distances. She calculated the heliocentric coordinates of each sunspot (its longitude and latitude on the sun's surface) and its size. She was fortunate to be engaged in interesting scientific work from the start of her time in Greenwich and even more fortunate that a famously large sunspot appeared in July 1892, less than a year after she began her new career. Its appearance was followed by a magnetic storm on the sun, also recorded in Greenwich.

The close working relationship between Annie Russell and Walter Maunder developed into something more and they married on 31 October 1895. He was forty-five, a widower with five children, ranging in age from twenty-one down to seven, and she was twenty-seven. According to civil service rules at the time, she had to resign her position immediately after her marriage but it was unthinkable that she should stop her work altogether. From this point on she became her husband's unofficial research partner, although a significant portion of her time would no doubt have been taken up with rearing his younger children.

The Maunders' butterfly diagram represented the culmination of decades of continuous observation of sunspots by astronomers in Greenwich and other observers. The findings were published in 1904, the same year the Maunders published other research showing that magnetic activity on the earth went in twenty-seven day cycles, linked with the rotation of the sun. The implication of this finding was that when there was a large magnetic storm on the sun, it would be directly facing the earth just one day in every twenty-seven of its rotation cycle. This was when the biggest impact would be felt on the earth. This is very important in the modern world because geomagnetic storms can cause serious interference in electrical grid systems, as well as satellite and ground-based communications. Annie Maunder analysed data at home after she resigned, while her husband made the daily sunspot observations. If anything, it seems that Annie was the dominant scientist in the partnership. In one of his books, Walter Maunder hinted at this when he referred to his wife as his 'helper in all things'.

Annie Maunder was an exceptional solar photographer and took better images of the very long coronal rays of the sun and solar eclipses using a small camera than many others did using larger instruments. She had built her camera using a grant of £40 from Girton. It had a lens of only 1.5 inches but had a wide enough focus to capture the outermost parts of the corona. The sun's corona is its outermost layer, which can vary and extend out millions of miles from the star's surface. The corona is somewhat mysterious: it is known that it is made up of intensely hot gas but the reason it is so hot has defied scientific explanation to date. It is two hundred times hotter than the visible surface of the sun, which is itself about 5500°C. Logic would suggest that the heat of the corona should fall off as it gets further away from the surface of the sun but the opposite is the case. Some scientists suspect that this reality may have to do with the nature of the sun's magnetic field.

Although she was conducting very high-level solar research,

Maunder was excluded from the professional body of astronomers, the Royal Society, because she was a woman. This meant that she could not publish in the society's journal under her own name. The result was that she and her husband became heavily involved in the British Association for Astronomy, a body that accepted women and amateur astronomers. Maunder became the editor of its journal, which was increasingly well respected under her editorship. The Maunders made trips to Norway (1896) and India (1898) to observe total eclipses of the sun. The Indian trip was the most successful scientifically and Maunder took some superb pictures, with her small camera, of the long tail of coronal rays, that spread out some ten million km from the sun. These were the first ever pictures taken showing the full extension of the coronal rays. The Maunders went on other solar eclipse trips, to Algiers (1900) and Mauritius (1901).

Maunder was also a leading expert in the field of ancient and archaic astronomy, a somewhat obscure area at the time but one that has become increasingly popular. She was interested in what people like the ancient Egyptians, Indians and New Zealand Maoris understood about the stars. She was also interested in the origin of planetary symbols and old astronomical texts, some of which were in the library in Greenwich.

Some of Maunder's own pictures were published in her popular science book, *The Heavens and Their Story* (1908). Although the book appeared with the names of both Maunders, Walter Maunder wrote in the preface that the book 'which stands in the joint names of my wife and myself, is almost wholly the work of my wife'.This book shows Maunder's skill as a popular science writer, interested in reaching a wide audience. It is written in an engaging style, does not assume readers' knowledge, draws on history, makes literary references and is nicely illustrated with photos and drawings.

Without a doubt, Maunder's most important legacy is her

work on the sun, its role in the earth's climate and its long-term periods of warming and cooling. This remains at the centre of scientific and political controversy today. There are those who believe that global warming – which all reputable scientists agree is now happening – is almost entirely due to the release of greenhouse gases (mainly carbon dioxide), which are a by-product of modern industrialisation and our way of life. There are others who believe the sun and its natural cycles are of far greater significance to the periodic warming and cooling of the earth. The current period of global warming began in 1950 or thereabouts, which some scientists believe is significant as this is when industrialisation began to intensify after the Second World War. Other scientists counter that there were many periods of global warming in the pre-industrial age that cannot be blamed on greenhouse gases, including the 'Medieval Maximum', which lasted from about 1100 to 1250. Their argument is that our current warming is just part of an ongoing natural cycle. The sun-inspired medieval warming went on for a hundred and fifty years so if we are in a similar period today, something that no one has either proved or disproved, the current warming might continue for many decades, then just stop. What is certain is that the work of Anne Maunder makes it impossible to rule out the role of the sun's cycles in global warming.

The Legacy of Annie Maunder

- She established the sun-earth climate link, a key factor in global warming.
- She proved, with the famous 'butterfly' diagram, that sunspots appear in predictable cycles.
- She was a superb solar photographer, who took landmark images of the sun's coronal rays.
- She was a role model for women in science, as one of the first ever salaried female astronomers.

Part II

TELEGRAPHS, STEAMSHIPS, SUBMARINES AND SPACE

5

TRANSATLANTIC CABLE HERO:

William Thomson, Lord Kelvin (1824-1907)

Birthplace: Belfast

In 1865, the year the American Civil War came to its bloody conclusion, the world record for the fastest crossing of the Atlantic by a passenger steam liner was held by the *Scotia*, of the British-owned Cunard Line. The *Scotia* had claimed the unofficial blue riband prize in July 1863 when it travelled the 5220 km (3243 miles) between Queenstown (now Cobh, County Cork) and New York in eight days and three hours, at an average speed of just under 27 km per hour.

By 1865, land-based telegraphs had been operating across much of the world for several decades. It was possible to connect cities separated by hundreds or even thousands of miles by sending electrical signals in Morse Code via wires held up by telegraph poles. The residents of Montreal and Miami, or of Paris and Vienna, could send and receive almost instantaneous telegraph messages, while those in New York and London could not. Even the historic news that General Robert E. Lee had surrendered to General Ulysses S. Grant in Appomattox, Virginia, on 9 April 1865 took more than a week to reach Europe by steam liner. The following year it would become possible, for the first time, to send messages across the Atlantic in milliseconds. Much of the credit for this engineering miracle must go to Belfast-born William Thomson, later known as Lord Kelvin.

In 1858 and 1865, employees of the Atlantic Telegraph Company had made brave but unsuccessful attempts to lay a submarine cable across the Atlantic. Then, in the autumn of 1866, the company finally succeeded in laying a submarine telegraph cable between Heart's Content, Newfoundland, and Valentia Island in County Kerry.

The name Kelvin resonates with us still, although William Thomson died in 1907. These days it is primarily associated with the Kelvin temperature scale, which we all learned about in school and which defines 'absolute zero' as -273°C, the coldest temperature that's possible in nature. This is also the temperature at which a system's entropy reaches its minimum value. A 'system' in this sense is anything that exists in nature, as a gas, liquid or solid, while 'entropy' is the term used to describe the level of physical disorder in the system. A highly ordered system is one where there is only one (or close to one) possible physical arrangement of its atoms and molecules. A perfectly formed ice crystal, for example, has an entropy close to zero, while a small amount of natural gas that has leaked into an office building with sealed windows has a high level of disorder or entropy, as there are many possible arrangements for its atoms and molecules.

Kelvin devised his famous temperature scale in an attempt to define in a more accurate way the precise nature of heat. The problem with existing temperature scales, he realised, was that they didn't reveal much about the energy in a given 'system'. For example, a solid, liquid and a gas might all exist at a temperature of 25°C but because their entropy differed, the levels of energy available from them to convert into work would be different. The solid would have the lowest entropy, because it is the most ordered, while the gas would have the highest, because it is the most disordered, while the liquid system would be in the middle. The solid would have greater potential to liberate energy than the liquid and more again than the gas. This is an important

'The Great Experimenter': Robert Boyle (1627-91), the 'father of Chemistry'.
(Portrait by John Riley c. 1689 © The Royal Society)

Buzz Aldrin of Apollo 11, who got to the moon with the help of Dublin-born mathematical genius William Rowan Hamilton (1805-65). (Photo courtesy of NASA)

Francis Beaufort (1774-1857), Chief Hydrographer (marine map-maker) to the British Admiralty. (Portrait by Stephen Pearce © National Portrait Gallery, London)

A watercolour of the grounds of Parsonstown (now Birr) Castle, showing the giant telescope, the Leviathan, built by William Parsons, the Third Earl of Rosse (1800-67). (Reproduction sanctioned by the Birr Scientific and Heritage Foundation, courtesy of the Earl of Rosse)

Human figures highlight the gigantic size of the Leviathan in Parsonstown. (Reproduction sanctioned by the Birr Scientific and Heritage Foundation, courtesy of the Earl of Rosse)

Photo by Alphonse Bernoud of the aftermath of the Great Neapolitan Earthquake of 1857. Bernoud worked for Robert Mallett (1810-81), the world's first seismologist. (Photo courtesy of Dublin Institute for Advanced Studies)

Nicholas Callan's induction coil.
(Photo courtesy of the National Science Museum, National University of Ireland Maynooth)

Nicholas Callan (1799-1864),
Professor of Moral Science in St Patrick's College, Maynooth.
(Photo courtesy of the National Science Museum,
National University of Ireland Maynooth)

Clare-born John Holland (1841-1914), technological visionary and inventor of the modern submarine.

John Tyndall (1820-93), Carlow-born Victorian man of science, lecturer, polemicist and defender of Charles Darwin's Origin of Species. *(Photo courtesy of the Tyndall National Institute, Cork)*

distinction. The Kelvin temperature scale was, therefore, a far better measure of potential energy than any existing scales.

Kelvin's interest in defining heat and energy in a better way led him to formulate the Second Law of Thermodynamics, 'thermodynamics' being the term used to describe the science that investigates the link between heat and other forms of energy. This famous scientific law states that heat will never flow from a colder body to a hotter body but will always move from a hotter body into a colder body. It is a simple enough concept. Think of a small radiator in a large room: heat flows from the radiator out around the room.

One of the other important things about Kelvin's Second Law was that it proved that heat energy from any high-temperature source – such as a steam engine – cannot be entirely converted into work – work, in this case, being defined as the energy required to drive the steam engine and move the locomotive forward. This was the first scientific explanation of why it is impossible to have a one-hundred-per-cent efficient heat engine. The law still holds today, as no one has yet managed to design an engine that is perfectly efficient.

This work on temperature was important, had an impact on the scientific community at the time and left a lasting legacy. However, it was something else that drew the attention of the public to Kelvin's genius. His worldwide fame and huge fortune were the direct result of his critical role in the laying of the first transatlantic telegraph cable, between Ireland and Newfoundland, in 1866. From this point on, almost instantaneous connections were available between Europe and North America. Businesses that were prepared to pay for telegrams, which were initially expensive, could gain a commercial time advantage over rivals. Important news no longer took more than a week to travel between the continents.

The man who would change the world was born in Belfast

in 1824, to a family of Scottish descent. Thomson's ancestors arrived in Ireland around 1641 and settled near the town of Ballynahinch, County Down, where they worked a farm called Annaghmore for almost two hundred years. His father, James, was born on the farm in 1786. James, who was self-taught, was scientifically gifted and managed, while still working on the farm, to attend the University of Glasgow, where he got an MA. This helped to land him a job teaching mathematics at the Royal Belfast Academical Institution (RBAI, more popularly known as 'Inst'), which is still a renowned school in the centre of Belfast. In 1817, James married Margaret Gardner, the daughter of a Glasgow businessman, and the couple had seven children. They lived in College Square East, next to the RBAI, and this is where Thomson was born, on 26 June 1824.

In 1830, when Thomson was just six, his mother died, leaving his father to rear seven children, ranging from twelve down to twelve months. In 1832 James accepted the offer of the chair of the Department of Mathematics in the University of Glasgow and Thomson and his siblings began their new life in Scotland.

Young Thomson and his brother James, who later also became an engineer, were allowed to attend their father's lectures. They built batteries and electrical machines and – remarkably – started their undergraduate studies aged ten and twelve respectively after they matriculated (passed the university entrance exam) in October 1834. Their age wasn't a problem, back then, and the boys regularly topped their classes and won prizes in physics and astronomy.

In April 1841, not yet twenty, Thomson headed for St Peter's College, Cambridge, after completing six college sessions in Glasgow but without completing his degree. In Cambridge, surrounded by gifted students, he flourished. Afterwards he went on to Paris for postgraduate studies but he always had an eye on heading back home to Glasgow. In 1846, opportunity knocked when the chair of the University of Glasgow's Department of Natural Philosophy (Physics) became vacant on the death of Professor William Meikleham. Immediately, Thomson and his father, James, began gathering testimonials to support his bid for the post. The young man was in a strong position, as his father was highly respected in Glasgow and his own talents had been recognised there. It was no surprise, therefore, that he was offered the job, which he held for more than fifty years. Thomson thus become the chair of the Department of Physics at the age of just twenty-two.

By the time Thomson reached his thirties he had conducted extensive research in his now famous Glasgow laboratory and made significant contributions to many areas of science. The young professor loved his academic work, spending time in his lab and teaching his students. His curious eye fell on the problems faced by the greatest scientific and engineering project of the mid-19th century – the transatlantic cable – and he was invited to join the project in 1856. From this point on it fascinated and consumed him, seriously reducing the time he spent on his other interests.

The Atlantic Telegraph Company had been set up in 1854 by an American businessman called Cyrus Field West. In 1837 Samuel Morse, an American inventor, had been the first to transmit a message successfully from one point to another, using electricity, via Morse Code. The commercial potential of being able to send almost instant messages between the big cities of North America was obvious to all: entrepreneurs, like West, got rich by building telegraph cable connections from Canada right down to the southern states and charging people to use the network. By the 1850s, West had set his sights on an even greater prize: connecting North America and Europe via a subsea cable.

West sought financial backing for his vision and set up the Atlantic Telegraph Company, along similar lines to his existing

American Telegraph Company. The world's first submarine cable linking England and France – a distance of some twenty miles – had been laid in 1850 but the Atlantic was vastly more challenging. It would require the skilled manufacture of more than 5000 km of cable at a time when submarine cable technology was primitive. It would also require a huge ocean-going ship, or ships, capable of carrying a monstrously heavy copper cable, and the input of some of the best engineering and electrical technicians in the world. Furthermore it would need the backing of wealthy investors.

Today, industry experts calculate that it costs about half a billion dollars to lay a new submarine transatlantic cable, without any extra specifications. Now many transatlantic cables connect, in milliseconds, US cities like New York and Chicago with London, Paris, Frankfurt and the rest of Europe (sometimes passing through Dublin). This service provides a tiny but crucial advantage to big financial houses trading on the stock market. Big financial hitters are prepared to pay a premium for this time advantage over rivals. The same was true in the middle of the 19th century and this possibly explains why, in 1858, West was easily able to raise £350,000 in risk capital to support his project, a significant but not enormous sum that would have had something like the same spending power as £20.5 million or €24 million today. The funds were raised from a relatively small number of shareholders who bought shares at £1000 each, a huge share price even by today's standards.

Most of the company's shareholders were British, with just one-twelfth of the funds raised in the US. Inside the British group, there was an influential group of Scots shareholders, who were keen to do all they could do to protect their investment in the project. For them, this meant ensuring that Thomson was elected to their board of directors. Now aged thirty-two, he had already proven his engineering and technical ability across many fields and the Scots considered him one of their own, as he had lived in Glasgow since the age of eight.

West was anxious to keep his Scots investors happy, so he asked Thomson to join the board of the company. When Thomson accepted his offer, he knew that the project would take him away from his other research and his beloved lab but the opportunity to get involved in the biggest engineering project ever initiated was too alluring. The judgement of the Scots cable backers was spot-on, as right from the start of the project serious technical difficulties arose that threatened its viability. The project was kept on track – barely – by Thomson's technical ability.

At first his role in the project involved doing little more than keeping an eye on things and making sure that the chief electrician, Wildman Whitehouse, and the other key technical officials were doing things correctly. He soon realised he was starting from a position of weakness as, by the time he came on board, many huge and irreversible technical decisions had been made. The most critical of these concerned the cable itself, which was to have a copper core through which the electrical signals would pass, surrounded by insulating material called gutta percha. All agreed on the materials to be used but the diameter of the cable was the subject of hot debate.

Gutta percha is the scientific name given to a family of trees native to several parts of the world, including southeast Asia, Australia and the Malay Peninsula. The same name is used for the natural latex, or rubber, that is produced from the sap of these trees. This latex was considered suitable for the transatlantic cable because it was not easily colonised or weakened by the activity of marine plants and animals, something that had been a problem for earlier subsea cables. Furthermore, gutta percha was known to keep its insulating properties even when a very high current was flowing through it. For these reasons, it became the insulating material of choice for underwater telegraph cables from the 1850s on. Eventually the overuse of gutta percha would lead to a collapse in the tree population supplying this natural rubber and the development of a superior synthetic insulating material called polyethlyene, commonly known as plastic.

The decision to use a light cable containing 107 pounds of copper 'core' per mile along with 261 pounds of gutta percha insulating material was made despite the opposition of Thomson and some other technical advisers to the project but there was significant pressure on the technical decision-makers, particularly Whitehouse, to keep the cost of materials under control. Thomson realised immediately that these kinds of cost-cutting decisions would ultimately lead to disaster. He understood how electricity moved through a subsea cable as he had done a lot of work in this area. In simple terms, he knew from his research that the only way electrical signals would successfully pass across more than 5000 km of ocean floor of varying topography was if the copper core and insulator were thick.

The contracts for the light core and insulator had been signed off before Thomson had a chance to influence the decision. After this, it was really a case of damage limitation from his point of view. From his own calculations, Thomson believed that for the copper core to be effective its weight must be 350-400 pounds per nautical mile, or three times as thick as the one ordered by the company. The same went for the gutta percha insulator: he understood that over a very long distance the electrical current in the subsea cable would be subject to resistance which would effectively block its passage, while there would also be a tendency for current to leak out of the cable and be lost to the Atlantic waters.

Thomson even invented a new law of physics, the 'law of squares', to explain and quantify this phenomenon. Its implications for the cable project were huge. For example, it showed that the level of disruption to an electrical current passing through a two-thousand-mile-long subsea cable would be a hundred times that of a cable two hundred miles long. This meant that the Atlantic cable was going to come up against far greater problems, in terms of resistance to current and leakage, than any other subsea cable. To counteract this problem, Thomson advocated that the diameter of the copper and gutta percha be increased in proportion to the length of the cable. The longer the cable, the wider it must be, or else the signals arriving at the receiving end would be useless.

It was only after the cable failed dramatically in 1858 that the company directors came to the conclusion that on technical matters they should have been listening to Thomson, not to Whitehouse. They blamed Whitehouse for applying excessive voltage to the cable to get it to work faster, 'frying' the cable and damaging it beyond repair. Thomson defended Whitehouse, although he hardly deserved it, emphasising the positive contribution he had made to the project from its inception. This didn't stop the company from firing Whitehouse. The first cable did work for a few weeks, during which time the press reported celebrations in the US, Britain and Ireland. In New York there was a hundred-gun salute and Irish cities reportedly witnessed 'scenes of wild delight'. The trigger for this public rejoicing was the successful transmission of the first transatlantic message by Queen Victoria, a ninety-nine-word telegram to US President James Buchanan.

What the public did not know at the time was that the Queen's short message – equivalent to a few brief paragraphs of text – took sixteen and a half hours to cross the ocean before being sent by land telegraph to President Buchanan in Washington DC. The telegram went from the new transatlantic telegraph station in Valentia Island, County Kerry, to the station in Heart's Content, Newfoundland; remote locations that had been chosen by the

Atlantic Telegraph Company because they were the two shortest landing points between the two continents. Neither would the public have known that it took just sixty-seven minutes to send exactly the same message from Heart's Content back to Valentia. The reason for the huge difference was that clerks deciphering the incoming messages in Valentia had been trained in the use of new, highly sensitive reading equipment designed by Thomson himself. Meanwhile, in Heart's Content, where Whitehouse held sway, the clerks tried to read the messages using older, less sensitive machines. This proved impossible and they began to use Thomson's readers but as they were not trained in their use at first, they could read messages only very slowly.

Whitehouse later lied to the board of the Atlantic Telegraph Company, telling them that the staff in Heart's Content had read the signals with his equipment. This and the cable-frying sealed his fate. He paid the price for the very public failure of the project and the resulting loss of public and investor confidence. The board gave Thomson complete control of engineering and technical matters from 1858, after Whitehouse was fired, and he brought the full weight of his scientific and engineering mind to bear on the cable project for the first time.

The first priority he identified was to assess properly the commercially available raw materials. It was known that pure copper was the best conductor, so Thomson began to analyse the levels of purity in the copper wire supplied by various manufacturers, finding that these varied significantly. He made sure he had the purest copper conducting material before he signed any more purchase contracts. Nothing was done in haste, unlike the first time around. Thomson was convinced that the project would succeed, having learned from the mistakes of 1857-8. His conviction and methodical determination drove the project forward.

Thomson faced enormous technical challenges. He had to

ensure that the cable for the next attempt was wide enough and its copper core pure enough to overcome the huge resistance to the transmission of electrical current that would build up over the transatlantic distances. He had to decide on the precise increase in the width of the gutta percha needed to hold the current on its long journey. There was also the problem of whether a boat existed that was capable of carrying and laying such a length of heavy cable. But the overwhelming problem was how to transmit electrical signals across the Atlantic and have clerks read them quickly and accurately. It would be very expensive, initially at least, to send a transatlantic telegram, so customers would expect speed and accuracy. Without these there would be no profits for the shareholders.

Thomson realised that even if he was able to improve greatly the quality and girth of the raw materials in his cable, the current sent from one side of the Atlantic would inevitably arrive at the other side in a significantly weakened state. He accepted this reality, whereas the mistake his predecessor Whitehouse had made was to try to amplify the electrical signal beyond the cable's capabilities. Instead of trying to boost the signal at source and risk frying the cable, Thomson set about building a new, very sensitive signal-reading apparatus at the other end.

The story goes that Thomson got the inspiration for his new super-sensitive reading apparatus when he absent-mindedly noticed light bouncing off his monocle while sitting one day in his office. His monocle was a single-lens instrument (one half of a pair of eyeglasses) that he used to correct short-sightedness in one eye and which he wore hanging from his neck. He noticed that when light hit the monocle it bounced onto the walls and reasoned that the precise location in which the light landed on the walls depended on the angle it hit the monocle. He wondered whether he could design a new instrument, using the deflective capacity of light, that would respond to tiny changes in electrical current over very long distances and provide the kind of reading accuracy he needed.

Thomson's subsequent invention, a mirror galvanometer, was ingenious and, more than any other technical development, paved the way for the success of the cable project. Silvanus Thompson, author of *The Life of Lord Kelvin* (1910) states: 'This mirror galvanometer, which with sundry modifications, was embodied in his [Thomson's] patent taken out February 20th, 1858, proved to be of enormous importance in the subsequent history and development of submarine telegraphy.' Without this instrument, many experts believe that the successful laying of a transatlantic cable might have been delayed for decades.

Thomson did not create the first mirror galvanometer, which was the invention of the German physicist, Johann Christian Poggendorff, in 1826. This galvanometer had a built-in mirror that moved in response to electrical current, reflecting a beam of light which was projected onto and read by a meter. It provided a very effective way of detecting the presence of electricity.

Thomson needed something that could pick up slight variations in current, as expressed in Morse Code, and detect even very weak signals, such as those that had travelled across the Atlantic. For this purpose he built his own mirror galvanometer that used a tiny mirror, with magnets on the back and an equally tiny piece of copper wire. When it was at rest, light reflected off the mirror onto a white screen or scale a few feet away – the zero position. When an electrical current was picked up through the tiny copper wire, the apparatus sprang into life, the magnets twisted in response to the current and a beam of light was reflected to one side or other of the point zero, to the right if the current was positive, to the left negative.

The electrical current that arrived in off the transatlantic cable passed through the mirror galvanometer, where the changes in light were read by a trained clerk, before the electrical signal went to earth. The changes in light patterns that were produced by the clerk on the far side of the Atlantic sending messages in Morse Code – which are pulses of electricity – could be translated into words. The more quickly this could be done the better, as speed had economic value. In 1858, the trained clerks in Valentia, using Thomson's galvanometer, were capable of reading the queen's message at the rate of about 1.5 words per minute. This might not sound impressive today but back then it was lightning speed compared to sending transatlantic messages by ship.

However, the cable failed in 1858: the signal tailed off over a number of weeks, before disappearing altogether. As Thomson had anticipated, it was too light for purpose. It snapped in two and it was not easy to put the two pieces back together again. The project stalled for a few years, then the company made another attempt in 1865. Again there were problems but Thomson, who was in charge, was able to overcome them. In 1866, for the first time, the world's biggest two financial centres, New York and London, were connected by cable. This date marked the birth of a global economic market for commodities, which could now be tracked as never before.

The British, who financed much of the cable project, stood to gain a lot, as the Atlantic cable was of huge value to them in communicating across their vast empire. The cable revolutionised global maritime commerce, which had been largely unchanged for thousands of years. Throughout history ships laden with goods went overseas and it might be years before the owner saw the ships again. There was no communication with the ship during this time and no way for business people to find out the price they had achieved, or the quantity and quality of the returning cargo. Maritime commerce was conducted largely in the dark before the cable came.

The change in the speed of communications between the old world and the new was staggering. International news organisations, powerful business interests and wealthy individuals could now send and receive information almost instantly, for a price. This was initially $5 per word at a rate of eight words per minute, far beyond the means of most people. Costs reduced steadily over the years, while speed improved. Today, there are transatlantic cables such as the Hibernia Atlantic Project Express which connects New York with London via fibre optic cable, offering just a five-millisecond time advantage to its customers. Times have changed but, in 2013, as in the 19th century, time means money.

For Thomson, the success of the cable meant that he became a very wealthy man and a celebrity in the US and Europe. He continued his pure research, following his own interests in areas such as thermodynamics and attempts to determine the age of the earth. He believed the earth was originally too hot to live on and that 'heat death' had occurred on the planet over its life span of twenty to four hundred million years. This view may have been coloured by his Christian beliefs, as the earth's actual age is 14.5 billion years. Thomson also refused to accept the theory of the gradual evolution of life as outlined by Charles Darwin in 1859 and clashed with supporters of Darwin.

Thomson wanted his research to be relevant and from early on in his career he proved excellent at spotting a business opportunity arising from it, anticipating the scientific entrepreneurs of the 20th century. He took out many patents, which contributed to his wealth. He bought a yacht, the *Lalla Rookh*, on which he spent a lot of his leisure time at his country mansion in Largs, a small, picturesque seaside town about 50 km west of Glasgow. His great wealth did not diminish his capacity to come up with inventions, particularly in relation to seafaring.

Thomson's father James had built astronomical dials by hand, so it was no surprise when his son built an astronomical clock in 1869. It had a number of dials showing the relative position of the sun, the planets and stars and was able to calculate how the various celestial bodies moved in relation to one another.

Thomson invented a mechanical sounding machine in 1876 that could measure to a depth of slightly over three miles. It lay just below a ship's waterline and meant an end to the old practice of taking soundings using ropes. The machine was worked by hand but could measure depth while the ship moved, unlike the ropes. Motor-driven depth sounders emerged in the early 20th century and these days ships mostly use echo soundings. They send out a signal, it bounces off the ocean bottom and is read by an on-board receiver.

In 1852 Thomson married Margaret Crum, whom he had known since childhood. His wife's health broke down soon after they were married and until her death seventeen years later, she suffered ill-health. The couple had no children. He married in 1874 for the second time, a woman called Frances Blandy.

Thomson was knighted by Queen Victoria in 1866 in recognition of his role that year in the laying of the transatlantic cable. He received a peerage on New Year's Day 1892, for his lifetime of achievement in science and technology. He was the first scientist to get a peerage for his work in science, although other scientists had held hereditary peerages.

When news came of the peerage, Thomson, as a commoner, had to choose a name that would be associated with his title and he opted for 'Lord Kelvin of Largs'. The name 'Kelvin' came from the River Kelvin, which winds its way through the west end of Glasgow, past the university, and was visible from his laboratory, while Largs was the coastal site of his primary residence. Kelvin is a Gaelic name, dating from the time when Gaelic was the everyday language of western Scotland.

These days, visitors to the beautiful University of Glasgow campus will see the Kelvin Building, which is home to the School of Physics and Astronomy. This structure in fact comprises two buildings, one constructed in 1905 and the other in the 1950s, so Kelvin himself never taught or did research there.

The Hunterian is the University of Glasgow's on-campus museum. It is home to several permanent displays, including one called 'Lord Kelvin: Revolutionary Scientist'. A collection of Thomson's scientific instruments is on display, along with diagrams and touchscreen displays, information about his life and personal artefacts, including some of his medals and papers and his pipe.

As Thomson got older, he spent more and more time in Largs, where he was an elder of St Columba's parish church. Faith informed his science and, while some scientists believed less in God the more they found out about nature, the more Thomson discovered about nature, the more he believed in one divine creator. He died in 1907, at the age of eighty-three. His funeral service took place at the University of Glasgow and he was interred in Westminster Abbey next to Sir Isaac Newton.

The Legacy of William Thomson, Lord Kelvin

- He was a key figure in the laying of the first transatlantic telegraph cable in 1866.
- He was a prolific marine inventor, of the marine compass, astronomical dial and sounding device.
- He formulated the Second Law of Thermodynamics.
- He was a role model for the scientist-entrepreneurs of the 20th and 21st century.

6

FULL STEAM AHEAD:

Charles Parsons (1854-1931)

Birthplace: Parsonstown (now Birr), County Offaly

On 26 June 1897, in Spithead, near Portsmouth, the British navy, the world's largest, is at anchor. The fleet has been assembled to salute the monarch, Queen Victoria, now aged seventy-nine, in recognition of her diamond jubilee (sixty years on the throne). Victoria's empire is at its peak, its tentacles stretching from Canada to New Zealand, to South Africa, to India. All this is made possible by naval power.

The show has begun: the guns fire, the bands play and colourful pennants flap in the breeze. Suddenly and without invitation, a small engine-powered yacht appears and darts between the neat rows of battleships, cruisers and destroyers. It is travelling at a speed of 34 knots, making it by far the fastest ship in the world. By way of comparison, the fastest ocean liner in the world at the time, the RMS *Lucania*, can reach no more than 22 knots, while the fastest destroyers in the British navy at the time can do a maximum of 27 knots.

The yacht weaves in and out of lines of warships, unimpeded, while the aged queen watches through a telescope from Castle Osborne on the nearby Isle of Wight. The yacht is the *Turbinia*; owned by Anglo-Irish aristocrat, inventor and engineer, Charles Parsons. Parsons has gatecrashed the queen's jubilee party to show the Admiralty bigwigs what even a modest boat can do when powered by one of his patented steam-turbine engines. The way he has chosen to do this is typically direct, innovative and effective.

The Admiralty and Parsons have some 'history' together. The County Offaly man invented his steam turbine in 1884. For much of the thirteen years since then he has been trying unsuccessfully to convince the Admiralty of the turbine's merits. He has arrived at Spithead, a naturally protected inlet off the English channel, for the jubilee day, with a plan born out of frustration and a growing sense of exasperation. He is desperate to show the naval elite that his steam turbine can help them make British warships faster, more fuel-efficient and far more destructive than enemy vessels, because they will have less need to carry heavy loads of coal and more space for big guns. The scheme he has come up with – which today would be regarded as a superb publicity stunt – is ingenious but it has its risks. Powerful figures will no doubt be annoyed by what they will see as a gauche display, performed purely for personal gain, on a day that was meant to honour their beloved queen, while others will see his invention a threat to their business interests.

Parsons would have known he had reached the point of no return as the *Turbinia* steamed through the perfect lines of the assembled fleet. He was steering the yacht, while his collaborator and old friend, George Stoney, stoked the engine madly below. The naval commanders present, shocked, gave the order to dispatch fast boats to intercept Parsons's craft but even the fastest naval boats couldn't get near the *Turbinia*. The message was clear for all to see. A small steam-powered yacht had outperformed the best the British navy had to offer. If the little *Turbinia* could do this, could an enemy warship not do far more with the same technology?

Parsons had forced the Admiralty's hand. The clouds of war were starting to gather again in Europe and it could no longer ignore such promising technology. Besides, Parsons might offer the engine to Britain's enemies, as his business instinct would trump any loyalty he felt to the Empire.

The Admiralty officers swallowed their pride and, despite being humiliated by Parsons and his *Turbinia* in Spithead, they made him an offer. The Parsons steam engine was to be incorporated into a new generation of British battleships, which became known as 'dreadnoughts', after the first of them, the HMS *Dreadnought,* was launched in 1906. This battleship made an enormous impact, with its speed and its array of big guns. Soon Britain's main rival, Germany, began building its own dreadnoughts. This led to a naval arms race between the UK and Germany, which continued right up to the outbreak of the First World War in 1914. A British dreadnought fleet met its German equivalent in battle just once during the war, in the Battle of Jutland off the Danish coast, but neither side secured a victory. In the Second World War, a type of 'super dreadnought' was also used until it became clear to everyone that these battleships had become sitting ducks for aircraft that could travel long distances to find and destroy them.

Parsons had created something new and outstanding, which, like many of the other great leaps forward in human history, had the potential to improve human life or destroy it. His engine certainly increased the capacity of the British navy and other navies to kill and destroy, changing the nature of war at sea, but it also revolutionised civilian sea transport, making it possible for people to travel long distances in greater comfort than before. The Parsons engine was the technology 'under the hood' during the glorious age of ocean liners. The Cunard line first fitted Parsons turbines into the RMS *Carmania* in 1905, later into liners like the RMS *Lusitania* and RMS *Mauretania* (both launched in 1907). These were fast ships that moved through the water beautifully, without the terrible vibrations associated with competing German passenger ships such as the *Deutschland*, which were powered by a quadruple expansion steam engine. In 1909, the *Mauretania* beat the *Lusitania*'s 1907 record for speed, taking the Blue Riband as the fastest passenger ship to cross the Atlantic, in a time of four days, ten hours and fifty-one minutes and holding this record until 1929.

The revolution in naval warfare and commercial sea travel that Parsons brought about by inventing the steam turbine was impressive but perhaps even more important to a much larger group of people was the fact that this new steam turbine could also be used to generate large quantities of electricity reliably. Up to this, electric lighting was primarily for the well heeled. Some streets in some cities were lit by electricity, usually in the most affluent areas, but most streets were still in darkness, or gas-lit. The turbines in use were poorly designed and noisy, as they had to work very hard to generate even small amounts of electricity. They were also unreliable. The Parsons engine meant that large sections of major cities, such as Chicago, had access for the first time to a reliable, cheap source of electricity.

Parsons was the youngest of four sons, the surviving children of thirteen born into the land-owning family based in what was then Parsonstown in County Offaly – Birr Castle is still the family seat. The Parsonses were different from other landed families, as they were interested in science, didn't rely on servants to do everything for them and enjoyed working with their hands in the estate's forge, experimenting with materials and making things. It was an extraordinary intellectual and scientific hothouse for a boy like Parsons and from an early age he shared his father's passion and talent for science. He got into trouble once when he built a toy helicopter powered by a little coal engine. The helicopter was capable of flight – remarkable, considering that the first full-scale helicopter did not fly until 1936 – but it crashed, coal fell on the floor and the carpet was burnt.

There was another early scientific adventure that went tragically wrong. Charles and his brother, Richard, built a car powered by a steam engine and drove it around the family estate. Their father's cousin, Mary Ward, came to visit. Mary was also a keen stargazer, as well as a good artist, and had an interest in insects. The boys offered Mary a ride in their car. She fell off, went under its wheels and died from her injuries. Thus, Mary Ward is recorded as the first person to die as a result of a motor vehicle accident and the tragedy of her death overshadowed the boys' achievement in building the car.

Parsons and his three brothers were encouraged by tutors chosen by their parents for their scientific ability. One was James Stewart, a well-regarded scientist in the field of mechanics in the University of Cambridge, whom Parsons heard lecture during his time there. Another was George Stoney, who became a lifelong collaborator and who accompanied him on his Spithead stunt. Stoney grew up in Oakley Park, County Offaly, only a few miles from the Parsonses' home. He became a research assistant to William Parsons and tutor to his sons.

It was no surprise that the boy said he wanted to study engineering in TCD. The young aristocrat spent two years in Trinity's engineering school, which had been established in 1840 in what is today the college's museum building. The Dublin university was way ahead of many of its UK counterparts in this area and the school produced engineers who helped to build the British empire, for instance designing railways in colonies like India and Burma. It was a university that had a reputation for producing hands-on engineers, students who knew how to take something apart, rebuild it, make improvements and understand how it all worked.

Parsons did not shine academically in TCD but began to do so in Cambridge, where he moved mid-way through his undergraduate engineering degree. In Cambridge he got a first class honours degree in mathematics. After he graduated he apprenticed himself to W.G. Armstrong, a major engineering firm in Newcastle, and his career took off. It was highly unusual for an aristocrat, the son of an earl, to work as an ordinary apprentice but he was happy to do so. He then moved to the Yorkshire-based firm of Kitsons, which was mainly involved in making munitions. During his time in Kitsons, Parsons worked on designing torpedoes, fired by compressed air. The next career move came in 1884, when, at the age of thirty, he moved to Clarke Chapman, a heavy-engineering firm based in Gateshead, close to Newcastle. The company's business was building engines for shipping as well as turbines that were capable of heavy-duty electricity generation.

In Clarke Chapman, Parsons thrived. It was during his first year with the company that he came up with his steam turbine, his greatest achievement and his lasting legacy to the world. In 1889, he decided to strike out on his own and founded C.A. Parsons and Co. in Newcastle, primarily to produce electricity generators in order to exploit the business potential of his revolutionary new design. He also set up the Newcastle and District Electric Lighting Company. Parsons, with his acute business mind, saw the potential his turbines had for lighting UK cities and towns that were still largely without electricity. After setting up on his own, he was worried about whether he had full ownership of the steam turbine design, which he had developed during his employment with Clarke Chapman. He tried to come up with a slightly different design, so that he could take out new patents over which he had certain control – a detour that proved unnecessary, as Clarke Chapman handed over most of the patents he wanted without any trouble. It was part of Parsons's personality to be painstaking and want to have control over all aspects of his life.

The first turbine designs created by Parsons were not very

efficient, generating only small amounts of power, but over the next fifteen years, he made small, steady design improvements. Already, in 1899, his first megawatt turbine engine had been built for an electricity generating plant in Elberfeld in the industrialised Ruhr area of Germany. Over time, as cities and towns bought his electricity-producing turbines, Parsons became immensely wealthy, reinvesting much of the wealth into his company. His engineering works expanded, producing some fifty turbines per year and providing many jobs in Newcastle. The company enjoyed success throughout most of the 20th century, reaching its zenith in the 1960s, when it employed some 7000 workers. It survives today as part of the power-generation division of the multi-national industrial giant Siemens, which bought the firm from Rolls-Royce in 1997.

What exactly is a steam turbine engine and how does it work? In simplest terms it is a machine designed to extract the thermal or heat energy that is present in pressurised steam. The energy is then used to turn a shaft around a fixed axis in what is called rotary motion. This is important, because rotary motion is a type that is well suited to the generation of electricity and the steam turbine engine had long been used as an electricity generator. The arrival of the Parsons engine represented a paradigm shift in the generation of electricity, as his turbine was the first capable of producing abundant, cheap electricity. His technological achievement can be gauged by the fact that today, one hundred and thirty years later, the vast bulk of the world's electricity is still produced by turbines that are essentially more advanced versions of the old Parsons turbine. There was no further paradigm shift in the intervening period

Up to Parsons's breakthrough, the problem with turbines was that they ran too fast for their own good and broke down too easily and too regularly, while the noise associated with them was horrendous, so much so that, in some cities, members of the public wanted them to be closed down. These engines were also incredibly inefficient in terms of the amount of fuel required to produce electricity, which meant that electricity in turn was expensive. Contemporaries of Parsons, such as the Swedish engineer Karl Gustaf Patrik de Laval (1845-1913), were also having some success in designing turbines that could generate a lot more revolutions per minute without breaking apart. But Parsons' great insights got him there first.

The problem for de Laval, Parsons and all the other engineers working on steam turbines was how to disperse the centrifugal force (the force drawing the rotating body away from the centre of rotation) that built up in steam engines. Then there was the problem of how to control the hugely damaging effects of high-pressure steam on turbine blades. Such steam could literally eat its way through the blades and destroy an engine. Many people had tried to tackle these problems but no one had succeeded in solving them.

Before Parsons, the best steam engines were capable of five hundred revolutions (the number of full rotations of a mechanical component around a fixed axis) per minute. In simple terms, the greater the number of revolutions, the more power or electricity the engine is capable of producing. For decades mechanical engineers had tried unsuccessfully to increase the number of revolutions in steam engines but, as the steam engine was a very complex piece of machinery with a lot of moving parts, any effort to increase the number of revolutions also increased the centrifugal forces acting on the engine, risking tearing it apart.

The beauty of what Parsons did was that he greatly simplified the steam engine and found a way to disperse the centrifugal forces and control the energy held in pressurised steam. This more stable, simpler, mechanical system could cope with a greatly increased number of revolutions per minute, creating a far more powerful and reliable engine. Almost overnight, with the Parsons engine, it became possible to generate 50,000 revolutions per minute, a colossal increase.

The way Parsons reduced the terrible centrifugal forces and the cutting power of steam was by creating a series of chambers within his turbine. The high-pressure steam would enter the first chamber and do some work, the pressure falling as a result. Then the steam, at reduced pressure, would enter another chamber where the same process would occur and the steam pressure would fall again. This meant that steam energy was exploited to the full but the steam was rendered safe by a gradual reduction in pressure. The Parsons engine could produce a lot more energy while working less hard than earlier turbines.

There was another problem to be tackled: the build-up of axial forces created by the constant pushing force on the shaft of the turbine, making it turn. Parsons created an equal and exactly opposite force of pushing, eliminating the axial load on the turbine. This idea came from a principle of balanced forces, which is a very important concept in design. In practical terms he removed the need for huge complex bearings at the end of the engine that carried the axial load. This was the pattern of Parsons's professional life: whenever he met a problem he came up with an innovation.

In personality Parsons was reserved and didn't suffer fools gladly. He was careful in everything he did, even the way he dressed himself every day. The evidence also suggests that he was happily married (to Katherine Bethell, 1883) and a loving father to his two children, Rachel, who later founded the Woman's Engineering Society, and Algernon George, called 'Tommy' by the family. The children seem to have spent a happy childhood in various Tyneside homes, close to their father's growing and successful business. It appears that Parsons developed some of his inventions, for instance a powered model helicopter and a three-wheeled go-cart, simply for the entertainment of his children. They were both keen on science and he, like his father before him, encouraged them by spending a lot of time working with them on various projects at home. They experimented on making turbine-powered machines, using methylated spirit as a fuel. They built a helicopter, as Parsons had done in his own youth, and something called a 'spyder', which sped around the room. There were echoes of Parsons's own upbringing when it started firing sparks and setting off small fires on the carpet. The children reportedly sped around the room, stamping gleefully on the smouldering embers.

When his son Tommy died in action in the First World War, at the age of thirty-one, Parsons was devastated. He had expected Tommy to return and take over the family business. At this point his daughter Rachel was running the company but Parsons held out on offering her a directorship, perhaps as a way of honouring his son's memory. Rachel resigned from the company as a result and the relationship between father and daughter broke down and was never repaired.

Parsons was a modest man, not the kind who would elbow his way to the front of a crowd, so his Spithead stunt was out of character and must have been born of intense frustration with the Admiralty which, he felt, was ignoring his valuable turbine invention. His political views would be regarded today as being profoundly right-wing but such views were common among the Anglo-Irish aristocracy at that time. He considered himself Irish in an Anglo-Irish Protestant way, typical of the time.

Though Parsons's turbine was by far his greatest achievement, it was one of many inventions, as his enquiring mind constantly sought answers to all kinds of problems. He was good at spotting business opportunities and, born wealthy, he gained greater wealth as a result of his talent and drive. He was his own man. For example, the design of one of his first steam turbines was literally drawn on the back of an envelope.

One of Parsons's weaknesses was a tendency to be obsessive, not knowing when to walk away from something. For example, he spent a huge chunk of his productive working life searching in vain for a way to produce artificial diamonds. In 1911 he presented a paper on the subject to the Royal Society in London and he is estimated to have spent at least forty years, 1882-1922, trying to reproduce some promising diamond experiments. His painstaking and methodical research continued until 1928, three years before his death, by which time he had spent much of his fortune searching for this Holy Grail. It was 1954 before Percy Bridgman produced the first commercial artificial diamonds.

Parsons was an intense, often difficult man, also capable of great loyalty. His best friend and engineering partner throughout his life was George Stoney. A great Irish scientist in his own right, Stoney invented the term 'electron', the fundamental particle that is responsible for the force we now call electricity. Stoney, who lived on the road in Dundrum, Dublin, that now bears his name and Parsons worked closely together for years, generating hundreds of patents and, although they had their ups and downs over the years, with Stoney breaking contact at one stage, they respected each other. The formal business relationship was that Parsons employed Stoney as his chief engineer but in reality they were partners in all engineering matters. The ability to bounce ideas off Stoney was one of the keys to Parsons's lifelong success,

Parsons died on board a steamship, *The Duchess of Richmond,* in Kingston, Jamaica, on 11 February 1931, at the age of seventy-seven. He and his wife were on a cruise of the West Indies. His place of death might be considered appropriate, given his huge contribution to making cruise ships faster, less noisy and less shaky, more fuel-efficient and more capable of reaching far-flung destinations. There isn't a mechanical or electrical engineering student alive today who isn't familiar with his work. People from Lima to Limerick switch on their lights and cook their dinners, thanks to electricity supplied by steam turbines based on Parsons's designs.

Garry Lyons, Adjunct Professor of Mechanical Engineering in TCD, believes that Parsons deserves to be considered on a par with the most famous Victorian engineer, Isambard Brunel, voted second only to Winston Churchill as a great Briton by BBC viewers in a 2002 poll. Yet how many people in Ireland have heard of Charles Parsons or his pioneering work?

The Legacy of Charles Parsons

- He invented a steam turbine which made sea travel faster and more comfortable.
- He provided the engine for the first dreadnought battleships, from 1905.
- He developed generators that produced abundant and reliable electricity.
- His engineering works spurred the growth of the city of Newcastle.
- He invented an early musical amplifier called the 'auxetophone'.

7

SUBMARINE WARFARE:

John Holland (1841-1914)

Birthplace: Liscannor, County Clare

It is 6 June 1878 on the banks of the Upper Passaic River in New Jersey, a short distance from the industrialised town of Paterson. A small band of Irish revolutionaries has gathered on the riverbank, including Jeremiah O'Donovan Rossa, a leading Fenian. They are about to witness something extraordinary: the launch of a 'sub marine' craft developed by an eccentric, thirty-seven-year-old teacher, John Philip Holland, from County Clare. A former Christian Brother and a recent emigrant to the US, Holland made contact with the Fenians through his brother Michael, who was a member of the secret society, soon after arriving in America.

After meeting Holland the Fenians, impressed by his knowledge and vision, decide to bankroll his submarine design. Today they have come to see if they have spent their money wisely. There is a lot at stake for them and for Holland and the air is filled with tension as they prepare for the field demonstration of the craft that will become known as the 'Holland I'. For Holland, a successful trial run will help ensure that his financial backers remained onside.

His boat is the culmination of his dreams of developing a submarine, which he has nurtured since he was in his late twenties and teaching in a CBS school in Dundalk. For the Fenians, this sub has the potential to be a powerful new weapon to threaten the all-conquering British navy. At best, a fleet of Holland subs could be developed to sink British warships blocking Irish ports following a rebellion against British rule. At worst, if only a single craft was built and deployed, it could be used as a weapon of psychological terror, silently targeting British warships, sinking them with its armed projectile, then disappearing.

When he arrived in the US, Holland used his Christian Brother contacts to secure a teaching position in the CBS school in Paterson. He is popular with his students, whom he regales with stories of mankind one day flying like a bird and navigating under the oceans. He draws designs for such machines on the blackboard during class and it's not clear whether his students think he is mad or a scientific genius. There is no doubt, however, that this is a passionate, confident and determined man. The school authorities are divided in their opinion of him, as he has little or no interest in maintaining classroom discipline or following the prescribed curriculum. He goes his own way, something that is not to everyone's taste. In his spare time and even sometimes during school hours, he is totally absorbed in creating designs for mechanical submarines and flying machines.

For Holland, the Fenians primarily represent a means of advancing his submarine ideas. On their first meeting, Holland showed them a two-and-a-half-foot-long wooden working model of his submarine and explained clearly to them how it would work. The Fenians present, who included the influential John Breslin, were impressed enough by Holland and his model to give him financial backing to develop a small boat. This became the 'Holland I'. It measured 14 feet 6 inches in length and 2 feet 6 inches in height, with a circular turret and hatch, and weighed in at 2¼ tons. It had a petroleum engine, a major technological advance on previous submarines, which had been powered by foot pedals. The Holland I had been constructed in great secrecy at an iron works in Albany Street in Lower Manhattan and moved to New Jersey for field trials. Today is the moment of truth. How will Holland and his craft perform when the pressure is on?

Holland prepares to dive underwater with his sub, which some have cynically dubbed a 'self-made coffin'. The boat disappears underwater and although no one at the surface can see it, the craft submerges to a depth of twelve feet and travels at just 3.5 mph. Holland has a great instinct for drama and has decided well in advance that he will put on a show for his important audience. He stays underwater for a long time, navigating in circles, knowing this will cause an atmosphere of drama and tension to build up on the riverbank. He finally surfaces a full hour after he disappears underwater and emerges from his hatch, smiling broadly. The Fenians on the bank are impressed and put in an order for a larger underwater boat, one that would be fully armed for combat and capable of sinking a British warship.

This was a pivotal moment in Holland's career, as it meant that he could give up teaching and go into designing submarines full-time. It was also a huge moment in the history of the submarine. Holland went on to design the world's first combat submarine for a naval service, the US navy, in 1900. He showed that he was made of different stuff from his Fenian brother when he sold submarine designs for Holland class submarines to the British navy between 1901 and 1903. He even built subs for the Imperial Japanese Navy, which ordered four Holland submarines in 1904 to use in the 1904-1905 Sino-Russian conflict over the countries' rival claims to Manchuria and Korea. Not bad for a lad whose childhood coincided with the Great Famine.

In the social context of the time in Ireland, John Holland was born into a relatively prosperous family in Liscannor, County Clare, in 1841. His father, also John, was a coastguard officer. This meant that he had a government salary and was entitled to live in a coastguard's cottage so the family's fate was not linked to the potato crop like many of their neighbours. The coastguard service had been an early warning system since the days when Napoleon had threatened the British Isles at the beginning of the 19th century and Holland's father spent much of his time going up and down the coast on horseback, watching out for enemy ships and smugglers.

John Holland senior's first wife was Anne Foley. She died in 1835 and he remarried, a woman called Mary Scanlon from Killaloe. They had four sons, including John. Mary was a native Irish speaker and John spoke Irish at home, learning English only after he started to attend St Macreehy's National School. He did well in primary school and attended the CBS in Ennistymon for a while. Then in 1853 the Holland family suffered a terrible blow when John senior died unexpectedly. Mother and sons moved to Limerick, where Holland attended the CBS in Sexton Street. All through his schooldays he excelled academically.

Holland joined the Christian Brothers in 1858 at the age of seventeen. It was a selfless move, done in order to remove the financial burden on his mother and secure his own future with a guaranteed teaching job. Brother Philip, as he was called in religious life, had shown himself to be excellent at science subjects and was sent for further training to the North Monastery in Cork. This school was home to Brother Dominic Burke, a famous science teacher, who was among the pioneers of technical education in Ireland. Brother Burke encouraged the shy but curious Holland to develop his talents. He also taught in CBS schools in Portlaoise (then Maryborough), Enniscorthy, Drogheda and Dundalk.

Holland was of middling height, 5 feet 8 inches or so tall, slender, with dark hair and an open, engaging personality. He was a gifted teacher of science and technology and an excellent draughtsman. He was very fond of music and his boys' choir in Dundalk CBS was well known all over Ireland.

The Christian Brothers kept meticulous reports, and in their annual *Vows Scrutiny Book*, Brother Philip didn't fare too well, according to the author, Richard Morris, in his authoritative book on Holland, *John P. Holland (1841-1914), Inventor of the Modern Submarine*. It was said that he was opinionated and unable to maintain classroom discipline and that he would rather dream of mechanical things than get down to the business of teaching the three Rs. Furthermore, the evaluation said that he resented being required to 'drill' the students, and, as a result, students in his classroom were relieved of such practices. Despite this official condemnation he was held in high esteem by many of his teaching colleagues and remained in touch with them long after he had emigrated to the US. He also supported the Christian Brothers financially and by letting them use his name as an endorsement when he became famous. He showed the same loyalty to new friends and supporters in the US, including William W. Kimball of the US navy in Brooklyn. Kimball said of the man who would become his lifelong friend that he was quiet, modest, informed and straightforward.

Holland's former pupils from across Ireland and in New Jersey remembered him as cheerful, happy, likeable, anti-establishment in nature and somewhat eccentric. They told some colourful stories about him, which ring true. One story relates that he built a mechanical duck while he was teaching in Dundalk. This duck looked very like a real duck and could walk about in the garden, swim, dive and surface again when put into water. In fact Holland appears to have spent much of his spare time during this period building mechanical devices, including a windmill, which pumped water from a well in the grounds to the top of the monastery. He also built a sundial and a wooden replica of the Rock of Cashel.

Physically, Holland was not strong. Even as a young man he suffered poor health and had infected lymph nodes, which caused sores on his neck. This ongoing health problem meant he was not permitted to take his final vows as a Christian Brother, although, in the end, this proved a good thing, as it meant he didn't have to break his vows when he decided to go to the US.

Holland's style of teaching remained the same when he got a job teaching in St John's Parochial School in Paterson, New Jersey. There, it seems, he regaled his students with colourful stories of futuristic mechanical wonders and flying machines. There is a mural in the Paterson Museum that shows Holland at his blackboard explaining mechanical flight with goose feathers attached to a broomstick, while on the blackboard there is a drawing of a one-man submarine. These were his two passions. The students were taken on a journey that was far away from the set curriculum, but many young minds were set alight and Holland's students remembered him years, even decades later.

He was a technological visionary, convinced, through his own knowledge of and familiarity with submarines, that the technology was sound and that it would change naval warfare. He was also convinced that 'heavier than air' flight was possible and outlined his approach in a paper called 'The Practicality of Mechanical Flight' which he produced in 1891. His crazy world of flying machines and submersible war vessels would become a reality within a few short decades.

It was while he was teaching in Dundalk CBS that Holland first began to think seriously about designing a submarine and drew up a sketch for a one-man iron submarine. He had read with fascination about the naval battles of the US Civil War, when iron-clad vessels first confronted each other. British leaders also noted these confrontations, fearful that iron vessels would in time become a major threat to their own wooden-bottomed ships.

His mother and brother decided to emigrate to the US in 1872 and the following year, Holland, again suffering from ill-health, left the Christian Brothers and sailed off to join them, carrying his first submarine sketch with him.

All the evidence suggests that Holland was far from being a

physical-force republican, although he had no great love of the British, especially after the way they dealt with the Famine in Ireland. But his first priority was finding a backer to support his projects. In his first few years in the US, Holland submitted plans to the US navy for consideration but they were rejected out of hand. Navy officials told him that it was not possible for anyone to go down in a submarine and that he'd be better off dropping his plans altogether. This drove Holland into the arms of the Fenians, who recognised the potential of his designs.

After the successful trial run of Holland I on the Passaic River, Holland began working for the Fenians on a three-man submarine that would be fully ready for war when complete. The Fenians allocated $20,000 for the building of the vessel, a huge sum of money in 1879, when the work began on the craft at DeLamater Iron Works in New York. The Fenians and Holland insisted on secrecy but news of a mysterious vessel being built in Manhattan leaked out to the city's press. Blakely Hall, a reporter with the *New York Sun*, wrote a report about the thirty-one-foot-vessel that could travel underwater and smash through a ship's hull. He called it the 'Fenian Ram', a nickname that stuck.

Holland designed the boat in the shape of a porpoise, to reduce drag, again using a petroleum engine. This innovation made the craft hydrodynamic so that it could travel at the same speed on and below the surface, about 9 mph. This was almost three times faster than the smaller boat he had demonstrated to the Fenian leaders on the Passaic. Designers of nuclear submarines after the Second World War imitated this porpoise-like shape.

Holland used sealed, compressed air to maintain balance and stability in the craft. There was a three-man crew: an operator, an engineer and a gunner. The gunner controlled a six-foot projectile, which could be fired by use of a pneumatic gun. The Ram could stay underwater for up to three days and could shoot its projectile in a straight line over a range of 300 yards. In a trial run in 1883 Holland reported that he reached a depth of 60 feet and was able to stay there for an hour. In the context of what had gone before in submarine design, the Ram was a technological triumph, a major step towards an operational combat submarine.

At last things were going well for Holland but just at that moment, as often happened in republican circles, there was a split in the Fenian leadership. As a result one group of Fenians took possession of the Ram and towed it to a 'safe' New Haven harbour, in an attempt to protect their investment. Holland, who had known nothing about this, was furious. He called it theft and vowed to have no more to do with the Fenians. He knew that they couldn't operate the craft properly without him and reportedly said, 'I'll let her rot on their hands.' The Ram was placed in storage and her engine removed and she did rot. She was not taken out of storage until decades later when, in 1916, she was put on display in Madison Square Garden, New York, to help raise funds for victims of the Easter Rising. Holland 3

The Holland 3 was an all-metal, sixteen-foot, one-ton replica of the Fenian Ram, which Holland built in the Gannon and Cooper yard in Jersey City after the Fenian breakaway group kidnapped the first Ram. The new, smaller Ram, launched in 1883, incorporated improvements that had been suggested in the trials of the original submarine

The Fenian connection was at an end and the big question for Holland was what he should do next. He was famous as a result of the press and public interest in the Fenian Ram. He had built up a good list of contacts but he was essentially broke. He took a job as a draughtsman in Roland's Iron Works in New York; then a gun inventor, Lieutenant Edmund Zalinski of the 5th US Artillery, offered him a job in the Pneumatic Gun Company. Zalinski wanted Holland to develop a submarine that would be equipped with his pneumatic guns. The two men agreed to work together, got investors on board and set up the Nautilus Submarine Boat Company as a joint venture. Holland then set himself to work on a new boat, his fourth, which became known as the Zalinski boat. Launched in Long Island in 1885, it was a fifty-five-foot long, cigar-shaped vessel but it was a technological failure and the Nautilus company was wound up.

Holland paused for breath for the first time in years and turned his attention to domestic matters. While teaching a choir in St John's School in Paterson, he had met Margaret Foley, an emigrant from Ireland. He was so impressed by her voice that he invited her to accompany his choir, although it was against the rules of the school. The couple were married in 1877, when he was forty-six and she was twenty-seven, and they had seven children. The family settled in Newark, New Jersey.

The impetus for Holland to get back in the saddle in earnest was the US navy's announcement in 1888 of an open competition for the design of a submarine torpedo boat. Among the specifications were that the boat should travel at 15 knots on the surface and 8 knots underwater; could remain underwater for two hours; could circle in a space no greater than four times its length; could dive to 150 feet in depth; could fire torpedoes with a hundred-pound charge of gun cotton. A whopping $2 million was to be made available to build the craft from the winning design and the competition was open to all comers.

There were four serious entrants and Holland was the winner. He was overjoyed but the navy rejected him because his building partner, Cramps Shipbuilding of Philadelphia, said he didn't meet all their specifications. No one else won either and another competition, on similar terms, was announced the following year. Again, Holland won and again he had bad luck, this time as a result of change in president. President Grover Cleveland had been keen on the competition but the new president, Benjamin Harrison, decided to redistribute the money Cleveland had allocated for the sub competition to fund surface boats that were already under construction.

This was very frustrating for Holland, who was again obliged to take a regular job, this time with the Cummings Dredging Company in New York. Charles Morris, the owner, was a supporter of Holland's going back to the Ram days. Then, in 1892, his luck changed dramatically yet again. Grover Cleveland was re-elected president and in March 1893 announced that he would reinstate the submarine design competition and make $200,000 available for the purpose.

In the spring of 1893 Holland met a talented, ambitious young attorney, E.B. Frost, and with him set up the John P. Holland Torpedo Boat Company in New York. As the navy had still not taken any definite steps towards commissioning a submarine, Holland and Frost began to seek opportunities to sell their submarine designs overseas. Navy officials were alarmed when they heard this and finally, in March 1895, they offered Holland's company a $200,000 government contract to build a torpedo boat. Holland decided that the new boat, his fifth, which would be called 'the Plunger', would be built in Baltimore. He moved to that city to oversee its construction. It was to be a massive craft, 85 feet long and 11½ feet in diameter. On the surface it would run on three steam engines, while an electric motor would power it underwater.

All appeared to be rosy in Holland's world until he began to feel cramped by the constant presence and interference of navy brass around him, fearing that the Plunger would be strangled by bureaucracy. In 1896 Holland's own company agreed that he would build a new submarine under a private contract. He would be free to design, build and test his craft as he saw fit. For Holland, his sixth sub, Holland VI, would represent the pinnacle of his career.

In 1898, at just the right time for Holland, war broke out

between the US and Spain over the issue of Cuban independence. The Holland VI, a superb machine, was built in Elizabethport, New Jersey. Holland borrowed from nature, devising a method of rapid diving similar to that used by many seabirds. The Holland VI was tested by navy personnel in November 1898, after the US had defeated Spain. They reported positively on some aspects but were of the opinion that the steering and diving were a problem, perhaps a consequence of the inexperience of the personnel trying to operate the boat. As a result the navy postponed a decision on whether or not to support the Holland VI submarine.

When he could not get the US navy to fund him to build a submarine, Holland reportedly said on one occasion: 'What will the navy require next, that my boat should climb a tree?'

In the winter of 1898-9, the Holland VI was given a total overhaul, a costly project funded by the magnate, Isaac Rice, an enthusiastic Holland supporter, who had provided batteries for the boat. In February 1899, Rice's Electric Boat Company was merged with the Holland Company. Rice became president of the merged companies and Holland was retained only as a manager. He never again had a major influence on the company and become increasingly isolated from this point onwards. The merged company installed a new skipper, Frank Cable, who trained under Holland for an official navy inspection of Holland VI on 5 November 1899. The boat passed the rigorous test with flying colours and the navy took possession of it for $160,000, making it the first official navy sub in the world.

Twenty-seven years after he landed in America and twenty-five years after he first approached the US navy, Holland, now aged fifty-nine, had finally had one of his submarine designs accepted by them. But he was being steadily sidelined by the Electric Boat Company and in March 1904 he resigned his position. He now tried unsuccessfully to get support for a high-speed submarine that could travel at 22 knots underwater, three times faster than any sub in the US navy.

Holland was brave and always performed his submarine trial runs himself, even when observers felt that he would drown inside his own invention. He was audacious and a master of the bold statement or unexpected deed. For example, during the 1898 Spanish-American war, he offered to take command of one of his subs and take a crew down to Cuba, where he would sink the Spanish fleet, if present, in the harbour of Santiago de Cuba. He said if the US government accepted his offer and he sank the Spanish fleet, he would expect them to buy his boat. The US government chose to ignore his flamboyant offer.

Most of all, he was a mechanical genius, who had received no formal education in engineering but had the vision, years before most other people, to see the future of the submarine and its potential impact in war. Writing on the subject of 'The Submarine Boat and Its Future' in 1900, Holland closely predicted the horrific damage German U-boats would wreak on Allied shipping in the Second World War, until the Allies cracked German codes and could track U boats and begin to hunt the hunters. Holland wrote: 'When the first submarine torpedo-boat goes into action, she will bring us face to face with the most puzzling problem ever met in warfare. She will present the unique spectacle, when used in attack, of a weapon against which there is no defence.'

Holland was a very determined person, which was shown by the way he persisted, when faced with a piece of bad luck that blocked his path towards success. According to Isaac Rice, one of his backers, he was 'never satisfied with what he invented and always seeking something new' – something that is true of all inventors. It is remarkable how many times in his life he achieved a spectacular professional success, only for something unexpected to happen that turned his success quickly into failure.

He certainly wasn't a man blessed by good fortune. Neither was he very good at business. His memory for business meetings was notoriously bad, perhaps because his mind was constantly preoccupied with issues of submarine design. On numerous occasions he was outmanoeuvred by lawyers and money men.

Holland's professional life ended on a sad note, as two of his former friends and supporters had taken over his company and patents and ensured that he could not set up a rival company. Even though he continued to have novel ideas and insights, he was unable to get financial backing for new projects, as potential backers feared legal complications.

As the years went by, Holland began to suffer from rheumatism and had a stroke which led to some paralysis in 1909. This interfered with his plans to build a flying machine in his barn and sadly it went to rust. He died in 1914.

The Legacy of John Holland

- He was the inventor and designer of the world's first combat submarine, which was purchased by the US navy in 1899.
- He was the first to invent a torpedo that could travel a long distance in a straight line.
- He designed and built the 'Fenian Ram' for the US-based Fenians in 1881. This was the first submarine to use a petroleum engine instead of pedal-pushing power.
- He designed the first subs for the British navy, the Holland-class submarines.
- He designed four submarines for the Imperial Japanese Navy in 1904, for which he received the Order of the Rising Sun from Emperor Hirohito.
- He was the first to design a submarine in the shape of a porpoise to reduce drag and make it possible for it to travel at the same speed above and below water.

8

MEN ON THE MOON:

William Rowan Hamilton (1805-65)

Birthplace: Dominic Street, Dublin

It can plausibly be argued that two men of Irish extraction, one born in Ireland and the other an Irish-American three generations removed, made possible mankind's greatest moment: the landing of two Apollo 11 astronauts on the moon on 20 July 1969. President John F. Kennedy, whose great-grandfather had emigrated from New Ross, County Wexford, to America a few years after the Great Famine, spoke on 25 May 1961 before a joint session of Congress. In what became one of his most famous speeches, he challenged his fellow Americans and most specifically the rocket scientists working at the National Aeronautics and Space Agency (NASA), to land men on the moon before the decade was out and return them safely to the earth. That, of course, is what happened, almost eight years later, before an awestruck global TV and radio audience of at least 600 million people.

Another Irishman, of a much earlier era, was vital to the success of the moon landings. This was William Rowan Hamilton, a Dubliner and mathematician. His role was critical because he had designed equations – now called quaternions – back in 1843, while walking on the banks of the Royal Canal, which gave the Apollo 11 mission controllers the ability to angle the returning Apollo capsule with great precision on its re-entry to the earth's atmosphere. If the angle of re-entry was miscalculated, it would result in certain death for the astronauts. These equations also helped the controllers guide the capsule, after the dangerous initial re-entry sequence, down through the earth's thick atmosphere towards its landing site.

Quaternions are still regularly used to land spacecraft, such as the Mars Science Laboratory (MSL) which landed on Mars in 2012. They are also the standard tool to guide and orientate satellites in orbit around the earth and they have been used to generate special effects in some of Hollywood's biggest budget films, for developing computer animations and for representing the ultra-realistic, relative movements of characters in 3D computer games.

It is unlikely that thirty-eight-year-old William Rowan Hamilton was thinking about future manned missions to the moon, as he walked with his wife Helen along the banks of the Royal Canal close to his home at Dunsink, on his way to a breakfast meeting in the city on the bright autumn morning of 16 October 1843. He was grappling with a problem that had vexed him for years: how to represent accurately and mathematically the relative movement of figures or objects in three-dimensional space, the space that we all inhabit.

For Hamilton, life was good that morning and about to get better. He was Ireland's Astronomer Royal, a job that gave him status with his scientific peers and provided him with a very nice house in Dunsink, next to the observatory, which still stands on a small hill a few miles north of Dublin. Though Hamilton's job at this time centred on astronomy, he was first and foremost a superb mathematician, dubbed by some 'the next Newton'.

It was a simple enough job, Hamilton knew, to represent two-dimensional space, like lines or points on a piece of paper. In 2D, there is an 'x' axis and a 'y' axis that intersect each other at a point called (0,0). All other points in this 2D world emanate from this, such as (1,3), which is 1 unit along the x axis and 3 along the y axis. That's the simple 2D world.

The representation of 3D is far more complex and no one in 1843 had come up with a good, or even half-way good way to represent points or objects that inhabited such 'real world' space. Hamilton later recalled that on this particular morning as he walked along the Royal Canal towpath, something quite extraordinary happened, akin to an 'electric spark' going off in his head. He suddenly realised that the solution to the problem was that there was no solution. The problem couldn't be solved in 3D; it could be solved only for numbers or objects in a theoretical 4D space and then applied back to 3D. He realised that four-dimensional numbers, a radical concept he later called 'quaternions', could give him everything he needed in 3D.

'His new numbers didn't satisfy the usual rules of arithmetic and they shocked not just mathematicians but society in general,' comments Dr Fiacre O'Cairbre, a mathematician in the National University of Ireland, Maynooth. O'Cairbre is an admirer of Hamilton, gives talks on his work and organises an annual walk along the Royal Canal to highlight the contribution of this rather unsung Irishman. 'His quaternions were very strange numbers at the time but they are now used all the time in computer animation, special effects and movie computer games,' says O'Cairbre. 'Quaternions are used in space navigation, in GPS, wherever you have to move things around, like on a computer screen, or in some kind of navigation.'

The morning Hamilton hit on the idea of 4D numbers, his immediate reaction was to record the insight before it evaporated. He surveyed his surroundings, took a penknife out of his jacket, of the type that most 19th-century gentlemen carried, and bolted towards a nearby stone bridge. There, in one of the great moments of mathematical history, he carved his quaternions into a stone block supporting Broom Bridge. The moment it became possible to land men on the moon and return them safely to earth thus had its origins on a quiet Dublin canal bank. O'Cairbre suggests that this moment could be described as 'one small scratch for man, one giant leap for mankind'.

This colossal breakthrough might never have happened, as Hamilton almost didn't make it past his nineteenth birthday. To find out why, we must wind the clock back from his moment of triumph in October 1843 to 17 August 1824. This was a huge day for Hamilton, a date he mentions in many letters. On this day he attended a party with his uncle, Reverend James Hamilton, at the home of the Disney family in Summerhill, County Meath.

Hamilton was born in 1805, the second of five children of Sarah and Archibald Hamilton. They were a well-connected family but their financial situation was not good, mainly because the family business had suffered from Archibald's connection with the failed United Irishman rebellion of 1798, which had shaken the British ruling class. Hamilton was sent, at the age of five, to live in Trim with his Uncle James, who ran a school of good reputation, which had been attended by notables like Arthur Wellesley, later Duke of Wellington. There the young Hamilton soon revealed his academic ability, especially for maths.

In 1823, Hamilton went to TCD to study maths and his brilliance was soon recognised. Before he even received his undergraduate degree, at the age of twenty-two, he was appointed 'Royal Astronomer of Ireland'. This was an important post that increased his profile. To return now to the Disney house party: there, the young mathematician was introduced to Catherine Disney, a girl of about his own age. According to reports, the two young people fell immediately and deeply in love. The following February Hamilton was apparently still in love: he wrote Catherine a Valentine ode and posted it in a letter. The return letter came not from Catherine herself but from her mother, bluntly stating that Catherine was to marry William

Barlow (a clergyman about twice her age). Hamilton was left stunned and heartbroken.

Hamilton was a young man of great ability with a bright future but he had three more years of undergraduate studies to complete in TCD. Perhaps, in the eyes of Catherine's family, he was still too far off securing a good, regular income. But such practical concerns probably never entered Hamilton's head as he was very idealistic and an avid reader of romantic poetry.

He took the news badly and, as he later described it, found himself standing alone on the bank of the Royal Canal, near the spot where he would triumphantly discover quaternions nineteen years later. He stared at the still waters, contemplating jumping in. The urge to leap into the water was strong, he recalled, but he hesitated as his mind began to rebel against 'the act'. Hamilton felt he had something to contribute to the world through science and mathematics, so he stepped back from the brink. Many people would benefit from this momentous decision.

Hamilton's beloved Catherine married the clergyman, in keeping with her family's wishes, and Hamilton also got married. But apparently neither of them stopped thinking about the other. In letters, Hamilton writes of his relationship with Catherine – or the lack of it – being like a violent torrent that ebbed and flowed throughout his life. It was a story well suited to Romantic poetry.

At this point, Hamilton's parents were dead, leaving him responsible for his four younger sisters, a considerable burden. Around this time he began to write poetry. His work was well received and he won several prizes for his poems, which were published in magazines. He saw no conflict between the 'emotional' aspect of poetry and the more 'calculating' nature of maths, describing the latter as 'an aesthetic creation, akin to poetry with its own mysteries and moments of profound revelation'.

Hamilton met William Wordsworth, by coincidence, when

they were both holidaying in England's Lake District. The two men took a walk together, became immersed in conversation on topics of mutual interest and become unshakeable friends. Hamilton was later godfather to Wordsworth's first child. Lady Jane Wilde also asked him to be godfather to her 'little pagan' son, the famous Oscar, and he was a life-long friend of the novelist, Maria Edgeworth.

It is July 1969 and hundreds of millions of people are watching and listening as Neil Armstrong sets foot on the moon. It is a monumental achievement but as Armstrong and Buzz Aldrin walk around the lunar surface, collecting soil samples and planting a US flag, they know that the most difficult and dangerous part of the mission is still to come: in particular the re-entry of the Apollo 11 capsule into the earth's thick and turbulent atmosphere. This is where quaternions came into play, as they gave NASA engineers a tool to orientate the capsule precisely for re-entry. The equations did this job superbly and enabled NASA to keep the capsule on track for its rendezvous with the USS *Hornet* in the Pacific. The capsule hit the water just fifteen miles from the recovery vessel, a spectacularly precise landing, especially given the late change of splashdown site due to unexpected bad weather.

Since the heady days of Apollo 11, these 19th-century equations have been the crucial pieces of mathematics used under the hood to navigate all kinds of spacecraft on all kinds of missions. On 6 August 2012, they were used to perform the most difficult spacecraft landing in history, when NASA landed its two-tonne Mars Science Laboratory (MSL) – about the same weight as the average American car – packed full of scientific equipment and apparatus, on the Martian surface. The MSL came in to land at a sharp angle, inside a crater called Gale Crater, next to an 18,000 feet tall mountain of layered rock called Aeolis Mons. This quite outrageously difficult landing site was chosen because it was where scientists believed there might be evidence for flowing water in the Martian past, or where fossils could show up in layered rocks. The landing would not have been possible without quaternions, which closely monitor the relationship between the spacecraft and its landing site, making tiny adjustments when required.

The engineering task facing NASA was to decelerate a car-sized craft travelling, as it entered the Martian atmosphere, at 13,000 miles per hour – much faster than a speeding bullet, which travels at about 1000 mph – down to 2 mph and land it safely and softly on a precise spot of Martian soil. It was a huge challenge and one of the key people involved in making it happen was Jody L. Davies, a flight mechanics engineer who ran many millions of simulations on a computer to try to predict all possible landing scenarios. But when it came to the real landing there was only one shot to get it right.

'The quaternions were part of what kept us on track,' said Davies. 'They [NASA] were aiming to get within a 20 km landing site; an order of magnitude less than anything before. This is the most precise landing of any craft on Mars. Gale Crater was desirable for them. We basically landed in a crater with a mountain in it. The mission is to see if Mars could ever have supported microbial life. You need the guidance to do it.'

Did she know about the origins of the person who invented the equations? 'I knew about Hamilton but had no idea he was Irish,' she said. 'It's a name that we see in our dynamics textbooks but I never put two and two together and realised he was from Ireland.'

The global computer gaming industry is now valued at €50 billion. As with space, quaternions have provided technicians with practical solutions to difficult gaming problems. Ten to fifteen years ago, the programmers who design games were looking for better ways to represent more realistically the movement of two or more objects rotating in 3D space. It was becoming more important, in this cut-throat business, to give players the feel of actually being 'in' the game, totally immersed in it, which required the characters to move like real people. Whether it was two warriors clashing on the Great Wall of China, or a platoon of US soldiers engaging the Viet Cong in Vietnam, the name of the game was ultra-realism. But what could provide it? In the 1980s Ken Shoemake, an academic based in the Computer Graphics Laboratory in the University of Pennsylvania, proposed quaternions to do the job. Programmers tried them out and soon found they worked perfectly.

To explain why, it is necessary to go back to the infancy of animation. In the early days of Walt Disney, animated movies were made by linking a long series of static frames in a continuous film that gave the viewer the illusion of movement. This was a laborious process as twenty-four static frames per second had to be drawn. A film of ninety minutes meant 90x60x24=a staggering 129,600 static frames.

These days, Pixar and other companies that make animated movies like *Toy Story* still use the standard approach of twenty-four static frames per second. But it is neither cost effective nor efficient to have the very best animators working on all these static frames, so they are employed to work only on the 'key frames', those that are most critical in representing a turning figure, moving from one position to the next. A key frame can be thought of as the midpoint between points A and B, where A is the starting position and B is the finishing position. There might be multiple static frames between A and B.

The creators of animated films use computers to identify the key frames, which are then drawn up by the animator. In the past, less talented animators were used to fill in the non-key frames; nowadays this job is also done by computer. It is the combination of the skill of the computer in identifying the key frames, and that of the animator in drawing these frames, that makes for an impression of very realistic movement.

The key frames are identified using quaternions. 'They are the perfect representation; the industry standard,' says Nick Gray, a computer games developer in the Dublin-based gaming multinational, Havok, which has developed special effects for Hollywood films like *Matrix Reloaded*. 'The problem of how to represent space was solved by Hamilton all those years ago. It's safe to say that any game in the last ten years with any animation in it has quaternions. *Halo*, *Battlefield*...all have quaternions. At last count, the *Halo* series had sold fifty million copies worldwide, while the extensive *Battlefield* series has sold a similar number of game units.'

Hamilton was hugely influential in many areas of his professional life but unhappy in his personal life. In 1833, he married Helen Bayley, with whom he had two children, but it appears that the marriage was more of a partnership than one based on love. The situation was not made easier by Helen's frequent illnesses of various sorts, while it is reported that Hamilton had periods when he drank excessively.

Catherine Disney had been obliged to marry her reverend but hers was not a happy marriage either. Late in life she decided to leave her husband and apparently she tried to reignite relations with Hamilton. It was too late but, when Catherine lay dying, Hamilton was one of the last to visit her and reportedly presented her, on his knees, with a copy of his book, *Lectures on Quaternions*. After her death, the letters Hamilton had written to her years earlier were found intact under her bed.

The Legacy of William Rowan Hamilton

- His quaternions were used to guide spacecraft, from the 1969 Apollo 11 mission to the Mars Science Laboratory landing in 2012.
- His quaternions are the standard tool for computer-game designers to depict ultra-realistic movement of figures in three dimensions.
- The orientation of all earth-orbiting satellites, which provide GPS and global communications, is based on his quaternions.
- Hamiltonian equations rewrote Newton's laws of motion, laying the foundations for quantum mechanics.

PART III

ATOMS, RADIO, PULSARS AND SPIRAL GALAXIES

9

ATOM-SPLITTING:

Ernest Walton (1903-95)

Birthplace: Dungarvan, County Waterford

He fires up his experimental apparatus, built of discarded glass tubes from petrol pumps and other odds and ends, and crawls across the laboratory floor. Ernest Walton, the twenty-nine-year-old Methodist minister's son from Dungarvan, keeps crawling until he reaches the safety of a small lead-roofed observation box on the other side of the room. This offers him protection from the huge electrical voltages and x-rays he expects to result from his experiment. Once inside the box, he sits upright to view his handiwork. This is the first time he has run this experiment, his first attempt to 'split the atom', and he isn't expecting much. But, to his astonishment, he immediately begins to see what he later described as 'twinkling stars – lots of them'. This is the tell-tale sign of the creation of alpha particles, the particles that Ernest Rutherford, Walton's boss in Cambridge University's Cavendish Laboratory, discovered years earlier. Rutherford predicted that alpha particles would appear only when the atom – or the atomic nucleus to be precise – was being split. That's what Walton is seeing. He phones his colleagues to tell them to get there fast. It is 14 April 1932. Scientific history is being made.

Rutherford and John Cockcroft, Walton's colleague, arrived hurriedly, in a state of high excitement. What happens next is pure drama. Rutherford, a large man, at this point aged sixty-two and not as physically flexible as he once was, is manoeuvred into the observation box by the two younger men. He looks at what is happening and begins to shout instructions: 'Turn up the voltage…Turn down the voltage.' The shouting stops. An unusually quiet Rutherford gets out of the box, walks silently across the room, perches himself on a stool and says, 'Those look mighty like alpha particles to me. I should know, as I was in at their birth.' The atomic age had begun, for better or worse, and a gifted young TCD graduate had created the experiment that made it happen.

It should be said at this point that although Walton is famous for having 'split the atom', this is not precisely what he and his colleagues did. In reality they did something even more difficult. They split the atomic nucleus, the tiny centre of the atom. There is a big difference between the two as the atom is far larger and easier to split than the tiny nucleus at its centre. The writer Brian Cathcart has compared the difference in size between the atom and its nucleus to the difference between a cathedral and a fly. Hence the title of his book on the subject, *The Fly in the Cathedral.*

On the historic day, Walton was trying out an experiment that he hadn't tried before. Cockcroft, along with the Russian scientist, George Gamov, had worked out that it might be possible to split the atom at far lower voltages than was hitherto thought. To everyone's amazement, the first time the experiment was tried, it worked.

One of the keys to the success of the experiment lay in Rutherford's decision to partner Walton with Cockcroft. In terms of personality and their approach to science, they were complete opposites. Walton was a superb hands-on scientist, who was capable of building any kind of scientific apparatus and designing experiments to suit a particular need, while Cockcroft was a theoretician, using mathematics to explore whether something could be achieved or not. He also had a reputation as an excellent 'scrounger', well capable of extracting both materials and equipment from commercial companies, an ability that was very important, as Cavendish was working on a shoestring budget compared to its US rivals.

The story goes that Rutherford, in the lead-up to the atom being split, had become impatient with the inability of his talented team to increase the electrical voltages to the one million or so volts that everyone thought would be required to accelerate charged particles to the point where they were capable of splitting the atom. The repulsive force around the atomic nucleus would naturally repel any other particle that tried to bang into it or go through it and split it apart. Only very fast-moving charged particles had a chance of getting through, something that would require huge voltage inputs.

This was all right down Walton's street, as he had done his PhD in an area that required the generation of extremely high voltages. But he still faced an enormous challenge and it seemed inevitable that the Americans, with their superior resources and equipment, would be the first to split the atomic nucleus. Already Walton, under pressure from Rutherford, had been risking life and limb to increase the voltages while working in a confined basement laboratory. The walls of the basement began to spark up with electricity and a decision was made to move above ground, to a new, safer laboratory, housed in a room with higher walls, where electricity would not pose so much of a threat.

The theoretical insights of Cockcroft and Gamov that led to a change in strategy by the Cavendish team were inspired by the emerging theory of quantum mechanics, which sought to explain the bizarre behaviour of many tiny particles found at atomic level that did not obey the rules of physics. One of the central concepts of quantum mechanics held that a particle could be positively and negatively charged at the same time, which broke all the rules. The importance of this for the Cavendish team was that in theory some particles, when accelerated towards the atomic nucleus, could tunnel through its repulsive barrier, while the majority flew up and over. A small number of accelerated particles could get through the barrier, a number sufficient to split the atomic nucleus.

There is a famous picture, albeit staged after the actual event, of Cockcroft and Gamov sharing the joy they felt when they discovered that the atomic nucleus was there for the taking. It could be split and this could be done at far lower voltages than was hitherto thought. The theoreticians had done their work splendidly and it was over to the team's nuts and bolts man, Walton, to show whether theoreticians were right about the lower voltages.

At Rutherford's urging, Walton had already managed to achieve voltages in the region of 700,000 to 800,000, so he knew that if the atomic nucleus could be split at a lower voltage than the one-million-volt target the experiment was unlikely to fail for lack of voltage. The Irishman now faced into the experiment that would change his life and mark one of the great moments in science.

The central strategy of the experiment – and every successful experiment must have a clearly-defined strategy – was to hit a piece of lithium metal, a light element, with positively charged particles called protons. The protons would smash the nucleii of some of the lithium atoms into pieces and create new particles in the process, called alpha particles. The charged proton particles would be created in the first instance by electricity interacting with hydrogen gas at the top of the experimental apparatus, which was maybe twenty feet tall, after which the protons would be accelerated downwards at massive speeds towards their lithium target. The lithium was surrounded by zinc sulphide screens, which would emit a flash of light if they were hit by alpha particles resulting from the smashing of atoms.

The moment of truth came and Walton got his apparatus going. As planned, the accelerated protons came roaring down through his apparatus towards the lithium metal target. As this was happening Walton scurried across the floor towards his observation box, for this was a dangerous experiment. To his wonderment he began to observe the little scintillations that were indicative of alpha particles hitting the zinc sulphide screens.

Walton, Cockcroft and Rutherford were delighted by the result and their unbridled joy is clear from the contemporary photographs of the men. They look quite exhausted and very relieved. Things began to move very fast. Rutherford, the laboratory leader, thought it would be prudent to secure the achievement and claim ownership of it as soon as possible, so he decided that the team would write a short article to be submitted to *Nature*, one of the world's leading scientific journals, describing what had been achieved. The normal procedure in scientific research is for a full article to be submitted, sent out for peer review, then returned to the author for clarifications and again reviewed – so it can take months, even years, before an article is published.

On this occasion there wasn't time for all that. The Cavendish team wrote a short letter, a classic of scientific understatement, and submitted it to *Nature* on 16 April 1932, two days after the experiment. The journal, which is published weekly, carried the letter two weeks later, on 30 April 1932. By the standards of scientific publishing, the letter had been published at lightning speed and everyone recognised its import.

The news that the atom had been split made world headlines and Walton and his colleagues found themselves at the centre of a media storm. Journalists, like everyone else, were amazed at what had been achieved but they were also asking tough questions about where this research on the atom and its huge potential to release energy might ultimately lead.

Wings over Europe, a play by Robert Nichols, was running in London when the news broke. By coincidence it includes a mad scientist who splits the atom, with dire consequences for mankind. Then along comes news that real scientists had split the atom. It is not surprising that the public's reaction was a mixture of awe, fear and unease about where all this might lead.

Albert Einstein was fascinated by the news emerging from the Cavendish Laboratory as the atom-splitting experiment had direct relevance to his own work. Einstein had predicted in his famous equation $e=mc^2$, where e=energy, m=mass and c=a constant, that energy and mass were linked and that a relatively small loss of mass could potentially result in an enormous release of energy. In other words, if a way could be found to break matter apart at the atomic level, Einstein believed it could liberate enormous amounts of energy. Here was an experiment that proved it could be done.

Einstein travelled to Cambridge to congratulate Walton and his colleagues in person for providing the proof he needed. There are crackly recordings of him talking about the importance of the atom-splitting experiment and it is clear that he regarded it as a major step forward for science.

Walton's life had been turned upside down and, still in his twenties, he was a famous man of science. Overnight, he became Ireland's most famous living scientist and he remained so until the day he died, in 1995, sixty-three years later. He never sought the limelight but many sought him out for advice and support.

After the momentous events in the Cavendish Lab, the world was his oyster and virtually any lab in the world would have welcomed him with open arms. But, true to his nature, Walton stuck to his original plan, returning to Ireland in 1934. His plan was simple: to work as a physicist in TCD, marry his childhood sweetheart, Freda, raise a family and to try to improve the standards of science and science education in Ireland. He managed to achieve all these goals in his long and contented lifetime.

For many scientists at the time – and this would be even more true today – his choice made no sense whatsoever and could only be seen as a backward career move. There was only a handful of third-level institutions in Ireland in 1934 and although TCD was the most famous of them, by international standards, it was very poorly resourced and staffed. The physics department that Walton returned to was something of a shambles. It had been neglected for years and it was not up to speed with new concepts such as the theory of quantum mechanics.

For Walton it was out of the question to continue with his research at the same level, or at any level. His primary role in the physics department would be to teach. He later became head of the department. In terms of research, he hit his high-water mark at a very early age and although he did not retire until 1974, forty years after returning to Ireland, he would never again approach the dizzy heights of scientific achievement he reached in 1934.

In some quarters Walton is associated with the development of atomic bombs, despite the fact that he was an active member of the anti-nuclear-weapons lobby group, Pugwash, in his later years. It could be argued that his role in the splitting of the atomic nucleus in 1932 was a significant step along the road towards Hiroshima and Nagasaki in 1945, but in reality many further scientific steps were needed before it was possible to make an atomic bomb.

For example, the Cavendish team used lithium metal atoms as their target for splitting, as lithium is a light element, relatively easy to split. However, for a nuclear bomb, it would be necessary to split much heavier elements – such as uranium – into pieces, then find a way to trigger a nuclear chain reaction, which would result in a truly massive release of energy. These developments would come within the next decade, driven by British and American fear of Nazi Germany developing its own bomb, but in 1932 they were still some distance off.

The Manhattan Project, to develop an atomic bomb, began in the US in October 1941 and was accelerated after the US declared war on Japan on 8 December of that year and Germany, as an ally of Japan, declared war on the US. The project hoovered up the cream of the world's research talent in the area of nuclear physics. Thousands of Americans began to work in several factories to support the project, without having any idea of what was behind it all. Many scientists were delighted to sign up for the project. For young and ambitious nuclear physicists, it was where the action was, where they could make a name for themselves. Scientists who had ethical reservations generally 'parked' those concerns and signed up because it was unthinkable that Nazi Germany would be the first to develop an A-bomb. Besides, the consequence of outright refusal was to be branded a traitor to one's country in its hour of greatest need and risk the total destruction of one's future scientific career. The pressure to join the project was immense in the context of the Second World War and very few scientists said no.

It was inevitable, given Walton's prominence and fame from the Cavendish days, that he would be approached to join the Manhattan Project. The approach came through an intermediary, James Chadwick. Chadwick knew Walton from his time in Cambridge and had won the Nobel Prize for Physics in 1935 for the discovery of the neutron, a particle in the atom that has no charge. Chadwick had agreed to take part in the Manhattan Project and became head of the British mission to the project, its chief recruiting officer. Chadwick asked Walton whether he would be willing to make a contribution to the Allied war effort but did not offer any details of the project.

Walton decided that ethically it was his duty to support the war effort and he approached the Provost of TCD to request permission to go to the US. The Provost at that time was an Englishman, William Thrift, who was also a professor of physics and very aware of the needs of the department. He refused to give Walton permission, saying that he had already lost his Head of Physics, Robert Ditchburn, to the war, and the college simply couldn't afford to lose another member of staff. Walton now had a clear choice. He could stay or he could hand in his resignation and head for the US, where his career would undoubtedly have taken flight. It was an easy choice for him, as he had no desire to leave Ireland. In later years he felt that he had been lucky at this crucial juncture in his life as the Provost's insistence that he stay gave him cover to make the decision he really wanted to make: to remain in Dublin with his family and continue working in TCD.

When he looked back on his career years later, Walton was very relieved that he had not taken part in the Manhattan Project, given what it led to. Though he declared himself opposed to the development and use of nuclear weapons through his work with Pugwash, in the extreme circumstances of the Second World War, he would have supported the use of any means necessary to defeat the evil of Hitler. He was not ethically opposed to what the people at the Project were doing but he was nevertheless relieved that he had not directly helped to kill people with A-bombs.

Walton was a rather quiet man, who didn't like to talk too much, but neither was he particularly shy. He was devoutly religious, a Methodist, a non-drinker and a very hard worker, spending long hours in TCD and in his office at home every day of the week but Sunday, which was set aside for religious services and rest.

The love of his life was his wife Freda (Winifreda Wilson), whom he met when they were both boarders at Methodist College in Belfast. He was the head boy, she the head girl, and everyone was aware of their affection for each other. They were both the children of Methodist ministers who had been posted to places throughout Ireland, so they had much in common. After school, they went their separate ways, Freda to teacher training college and Walton on a scholarship to TCD. A few years later Freda took up her first teaching post, in Waterford, and Walton achieved his ambition by being accepted to do a PhD at the Cavendish Lab.

That would most likely have been that but for a chance meeting on a Belfast-Dublin train that broke down near Newry. On the platform, Walton and Freda spotted each other. The romance was quickly rekindled and a torrent of letters began to fly between Cambridge and Waterford, letters that speak of their great love. All Walton's letters to Freda were donated to TCD and are now held in the library there. He wrote to Freda on the day the atom was split to say it had been a 'red letter' day for two reasons: one because they had split the atom and secondly because he had got a letter from her. After he returned to Ireland with the world at his feet, he married his beloved in the Methodist church, just off St Stephen's Green. All that remains of the building now is the façade as it was destroyed in a fire.

Walton had a dry sense of humour but in general he kept his thoughts to himself and even his family didn't know his views on many matters. He was very much a traditional Methodist, with a huge commitment to his job and his role as an educator. Although he was devoutly religious, his son Philip recalls that he wouldn't talk about his faith at all, unless, for example, asked to give a lecture on science and religion. Neither would he talk about his work, or refer to the fact that he was so highly thought of in the science world.

One of Walton's closest friends was a Professor Cocker, who was a Methodist who had come from England, and whom Walton took it upon himself to look after. They occasionally went on holidays together but even this man, his best friend, had no idea that, in 1951, Walton was about to be awarded the Nobel Prize for science. Walton knew he was up for the award but simply didn't mention it. He put a great amount of work into preparing his lectures: scientists working in universities today might be shocked to learn that he delivered up to fifteen lectures every week. He was also a master of the lecture demonstration, when he would have an apparatus out on the bench, doing an experiment while giving his lecture. This greatly helped Walton's students to grasp the material he was teaching.

Walton's son Philip remarks that a Nobel laureate in the US would probably not have to do any teaching at all, then or now. He would be a prized asset for any university and every effort would be made to maximise his research talent. Walton was Head of the Physics Department in TCD for many years but operated without any secretarial back-up, doing much of the department's correspondence himself on an old typewriter in his home in Dartry. When his children were young he worked in the upper reaches of the house, almost in the attic, and as the noise subsided and he got older, he gradually worked his way down to the ground floor.

His daily routine was to get home from work, have his dinner with his family, then disappear upstairs to his office. This wasn't a lot of fun for Freda, who occasionally tried to coax Walton to stay downstairs by saying that there was something on television he might like to see. The best she could hope for was that he would come to the door of the television room and stand and watch something for a little while but he was always anxious to escape back to his office as soon as the opportunity arose.

He is fondly remembered by scientific colleagues and former students as an excellent teacher, a tremendous experimental scientist and a modest, very nice man. He remains Ireland's only Nobel laureate in science.

The Legacy of Ernest Walton

- He was part of the scientific team that split the atom in 1932.
- His atom-splitting apparatus was the forerunner of today's huge atom-smashing machines, such as the Large Hadron Collider, which is located underground at the Franco-Swiss border.
- He was a strong advocate of the peaceful use of nuclear power.

10

THE WIRELESS WIZARD

Guglielmo Marconi (1874-1937)

Marconi's mother Annie Jameson, was born in Enniscorthy, County Wexford

At 11:40pm on 14 April 1912, the *Titanic* hit an iceberg. There were two Marconi wireless radio operators on board, Harold Bride and the senior operator, Jack Phillips. These men didn't work for the White Star Line that owned the *Titanic* but were employed directly by the Marconi International Marine Communication Company. This had been set up by Guglielmo Marconi, the thirty-eight-year-old, half-Irish, half-Italian wireless communication wizard and joint winner of the 1909 Nobel Prize in Physics. Phillips, twenty-five, was from Surrey and had received his wireless training in the Marconi transatlantic wireless station in Clifden, County Galway, before being put to work on the big ocean liners.

On the night the *Titanic* sank, Cyril Evans, the Marconi wireless operator on the SS *Californian*, located a few miles from *Titanic*, had sent Phillips a warning of pack ice in the area. Phillips was in the middle of trying to establish contact with the wireless station in Cape Race, Newfoundland. He was irritated by Evans's message because it was very loud, given the proximity of the *Californian*, and it had interrupted his work. In Morse Code, Phillips bluntly said to Evans, 'Shut up, I'm working Cape Race.' Fifteen minutes later, the *Titanic* hit the iceberg. After the big collision, Phillips began sending out wireless distress calls in all directions. The *Californian* was the best placed to respond but Evans had by then turned off his wireless set and gone to sleep for the night, so his ship sat idly by as the *Titanic* sank at 2:20 am.

The news of the *Titanic*'s fate did not reach Evans until he turned on his wireless set at 5:30 the next morning. The shocking news was first transmitted by the RMS *Carpathia*, which had wireless on board. It responded to the distress calls although it did not arrive on the scene until 4.00am as it had been several miles away. The *Carpathia* managed to pull more than seven hundred people from lifeboats and the sea.

Distraught relatives gathered at the offices of the White Star Line in New York the next morning. The *Carpathia* transmitted the names of the survivors there via Cape Race. In the aftermath of the *Titanic* disaster, the life-saving role played by wireless technology in the tragedy was reported in the press, as was the name of the man behind it all, Marconi. This led to global fame for Marconi and the first widespread realisation of the huge power of wireless technology.

When the *Carpathia* arrived with the survivors in New York, Marconi was there on the dock to meet the ship. He went on board to talk to the survivors, along with a reporter from *The New York Times*. He spoke to Harold Bride, the surviving Marconi wireless operator – Phillips had perished in the North Atlantic while clinging to a lifeboat. A few months later, on 18 June 1912, Marconi appeared before a US court of inquiry, set up to investigate why the *Titanic* had sunk and what could be done to avoid a similar tragedy in future. The findings led the US Congress to pass the Radio Act of 1912. This stated that all passenger ships must be equipped with wireless radio communications, that wireless must operate around the clock, that radio contact must be maintained with ships in the immediate area, as well as the coastal radio stations, and that a back-up power supply must be in place to ensure that wireless was working at all times. For Marconi this was good news, as it meant that more vessels on the high seas would be requiring his technology and his operators. The tragedy of the *Titanic* thus resulted in personal triumph for Marconi. He was lucky for another reason. His daughter Degna later revealed that they had been offered free passage on the *Titantic*'s maiden voyage but had decided instead to travel to the US on the *Lusitania* three days earlier. Marconi had a lot of paper work to do and there was a stenographer on the *Lusitania* but not on *Titanic*.

The name of Marconi, father of radio, is revered in Italy, where many museums and societies have been set up in his honour. He was, however, as much Irish as he was Italian and it could be argued that Ireland has a stronger claim on him than Italy. Marconi was the second son of Giuseppi Marconi, a well-off landowner from near Bologna, and an Irishwoman, Annie Jameson, a member of the Jameson distilling dynasty. Annie met her husband in Italy when she went there to train as an opera singer, against her parents' wishes. She was nineteen when they met but she and Giuseppi did not marry until she was twenty-five because of her family's opposition.

Perhaps as a result of this experience, Annie wanted her son to be able to follow his passions freely. When the young Marconi began to tinker with radio experiments in the garden of the family home, Villa Grifoni, it was Annie who provided him with emotional support when his father considered what his son was doing a waste of time. It was Annie who provided Marconi with a list of contacts in England and Ireland, people who would prove crucial to the development of his career, when he found it hard to find financial backers and make progress in Italy. She introduced her son to a group of wealthy Irish grain merchants who supported him in the early days when he needed funds to build his very large radio transmitters.

There was another key Irish connection in Marconi's life: his first wife, Beatrice O'Brien, was the daughter of Edward O'Brien, 14th Baron Inchiquin, whose seat was Dromoland Castle. Edward died in 1900, when 'Bea' was eighteen, and over the following years she and her mother were regulars at social events in London, where society people found suitable wives and husbands. Bea met Marconi at one of these gatherings in 1904, when he was working at a wireless station at Poole in southern England. Marconi was attracted to Bea and, even though he had been involved with a young American woman called Inez Milholland, he proposed to her. She initially declined: it seems she was quite intimidated by his celebrity status as he had been famous for his wireless work since the late 1890s.

True to his character, Marconi persisted until Bea agreed to marry him, despite objections from her family because of Marconi's half-Italian origins. They were married in St George's Church, London, on 16 March 1905 and plenty of onlookers and press were there to see the ceremony. The couple honeymooned in Dromoland Castle but Marconi had to cut their honeymoon short to return to London on business. This was a sign of things to come, as such absences later contributed to their separating.

Bea and Marconi had four children: three girls, Lucia (who lived only a few weeks), Degna and Gioia, and a son, Giulio. The family lived in Southampton for a time, before moving to Rome in 1913. There Bea enjoyed the social life of Italian high society. During the First World War, Marconi was at home more than before and all was well. However, after the war ended and he began again to travel extensively on business, the marriage came under strain. There were rumours that he was involved with other women and the couple began to live separate lives. In 1923, it was Bea who requested a divorce, as she had fallen in love with Liborio Marignoli, Marchese (Marquis) di Montecorona, whom she later married. It seems, however, that Marconi remained bitter in the aftermath of the divorce and Bea never saw him again after 1928. When he died in 1937, he left nothing in his will to Bea or their three surviving children. Bea went to see him as he lay in state, unnoticed as she mingled with thousands of mourners.

It was in Ireland that Marconi carried out some of his most important work, which far outweighed anything he did in Italy. For example, the 225-mile 'proof of concept' experiment – which was set up to test whether radio waves would follow the curvature of the earth as Marconi suspected – took place between Poldhu in Cornwall, his wireless transmitting station, and Crookhaven in west Cork, one of his receiving stations. This set the scene for the famous experiment in 1901, where Marconi sent the first wireless signal across the Atlantic, between Poldhu and his station in St John's in Newfoundland.

Throughout the 19th century Crookhaven was the first safe port for shipping arriving into Europe from North America. There were established methods of communicating with ships using flags and lights, but when wireless came, the arriving ships communicated directly with Crookhaven via wireless. This news was very important for the markets in London as these first communications from New York helped the city of London to track where goods and commodities were located and to set prices to match their arrival times in London or other British ports. Crookhaven was also home to a Reuters news bureau. Before Marconi, Reuters sent reporters out to arriving ships to get the latest news and send it with the reporters on a train to Cork, where it was relayed to London via telegraphy through a fixed cable. Marconi speeded up the transatlantic transmission of news by several hours by sending instant wireless messages from Crookhaven to Cork.

It was on Rathlin Island, off the coast of County Antrim, that Marconi set up a wireless station in 1898. This station, which was linked with Ballycastle, provided wireless monitoring of shipping passing between Ireland and Scotland, a service specifically for Lloyd's insurance company, which wanted to be able to track ships in this area in bad weather, something they couldn't always do using line of sight from the mainland. From Rathlin it was possible to see ships even in heavy fog and relay the information wirelessly to Ballycastle and from there to the offices of Lloyd's. This was valuable information for the company, as a major insurer of ships. Rathlin was important for Marconi as this arrangement with Lloyd's was his first commercial wireless contract.

Also, in 1898, Marconi provided the first wireless coverage of a sporting event, the regatta that took place in Kingstown (now Dún Laoghaire), five miles south of Dublin. The *Dublin Express* had commissioned Marconi to send wireless reports to its editorial offices about the position of the yachts, so Marconi set up his wireless transmitter on board a boat called *The Flying Huntress* and his receiving equipment in the newspaper's offices. Marconi sent in the region of seven hundred messages and the *Dublin Express* ran extra pages providing the most up-to-date reports on the regatta.

In 1905, Marconi began to look for a site for a big new wireless transmitting station that would replace Poldhu. He had his eye on the transatlantic market, so it made sense to look in the west of Ireland, the nearest point in Europe to North America. For a number of reasons he settled on a site of just under three hundred acres a few miles south of Clifden. There was a direct line of sight across to his receiving station in Glace Bay, Nova Scotia. There was an abundance of good quality turf available locally to feed the electrical generators needed to transmit messages and keep equipment cool. Finally, the site was only eighty miles or so from Dromoland Castle, his wife's home place. Marconi's connection to the Inchiquins was important to him as it meant that he would not be treated as an outsider by the locals.

The station in Clifden began operating in 1907 and was a huge commercial success. Marconi could provide thirty words per minute, instead of the standard twenty-two per minute provided by the cable telegraph companies, and the costs of transmission were half what the cable companies charged. Marconi was in direct competition with these companies but he was smart enough to use cable when it suited him, such as laying a permanent fixed cable between Clifden and Galway, which was in turn connected to London. On the other side of the pond a fixed line connected Glace Bay to New York via Montreal. This meant virtually instantaneous communication between London and New York.

The transmitting station employed two hundred people at its peak and threatened the iron grip the telegraph companies had had on transatlantic messages ever since a reliable subsea transatlantic cable was laid in 1866 in a project led by another Irishman, William Thomson, Lord Kelvin [*see* Chapter 5]. Wireless was another huge step as communication was now even faster than through fixed cables.

The transmitting masts at Clifden were huge, about two hundred feet in height, which is more than half the height of the RTÉ mast in Donnybrook today. They required a lot of metal, solid wood and other materials as well as labour, the supply of which greatly helped the local economy. Marconi wanted to produce low-frequency radio waves, or long-wave radio, the type of radio waves that are capable of travelling long distances, bouncing off the earth's atmosphere in the process. Very large capacitors – a capacitor comprises two metal plates between which an electrical field is stored – were required for such transmitters.

In June 1919 Clifden and Marconi were in the headlines again when Alcock and Brown's modified First World War bomber crash-landed in a bog next to the wireless station. The men arrived just a few minutes short of sixteen hours after their take-off in St John's, Newfoundland. It was no accident that they landed in Clifden (although it was certainly not their attention to crash-land) as the wireless station there meant that news of their arrival in Europe could be instantly conveyed to the offices of the *Daily Mail* in London. The newspaper had offered a prize of £10,000 to 'the aviator who shall first cross the Atlantic in an aeroplane in flight from any point in the United States of America, Canada or Newfoundland and any point in Great Britain or Ireland in seventy-two continuous hours'. With their eyes on the valuable prize money, Alcock and Brown reasoned that if they landed at Clifden and sent the news on to London immediately, they could claim the cash even if another plane was physically in the air ahead of them somewhere. It was a clever tactic but they needn't have worried as theirs was the only plane in the air at the time. For Marconi, the link between the pioneering flight of Alcock and Brown and his wireless station brought more valuable commercial publicity.

Marconi should not truly be thought of as a scientist, or even as an inventor. Surprisingly, although he was technically very gifted, he invented little or nothing that could be classified as new. His genius lay in business: being able to spot the value of new knowledge wherever he came across it and exploit it for the maximum commercial benefit. He wasn't university trained, although he had received private tuition from a professor of physics in Bologna. He was self-taught to a large degree and from an early age was reading popular electronics magazines that explained how to do experiments.

One of the big criticisms Marconi faced throughout his career was that he was simply borrowing or stealing other people's ideas. This criticism reached a head after he won the 1909 Nobel Prize for Physics for his work in radio. Many people were upset that

Marconi got this prize, as others had contributed genuinely new knowledge that had helped the development of radio. One of these was Oliver Lodge, a British scientist who believed that he had been the first to demonstrate wireless communication in 1894, several years before Marconi. Lodge was correct but for him, a scientist, it was enough to do it and move on. However, on a business level the demonstration of the technology was only a first step and it was Marconi who moved it onto the next level. His talent lay not in making breakthroughs but in adapting technologies, from whatever source, into something useful.

Lodge was not the only person unhappy with Marconi getting all the credit for radio. Perhaps the most famous dispute surrounding the issue of who invented radio was between Marconi and the Serbian-American inventor and scientist Nikola Tesla, who developed the AC, or alternating current system of electricity. Both men applied to the US patent office as the inventor of radio and Tesla was granted the patent in 1900. Marconi lost this battle but his determination ensured he eventually won the radio war.

Aside from wireless, Marconi also features strong in the history of the development of radar. This technology was developed by the British at the start of World War II and gave the RAF (Royal Air Force) a critical technological advantage over the Luftwaffe (German air force), which helped it win the Battle of Britain in 1940, despite the fact that its planes were outnumbered by about three to one. In 1922, eighteen years before the war in the air, Marconi gave a talk to the Italian Institute of Radio Engineers, at which he expounded his ideas for a radar-like technology. At that time, he was thinking in terms of the technology being used to guide ships at sea. He said, 'It seems to me that it should be possible to design apparatus by means of which a ship could radiate or project a divergent beam of these waves in any desired direction, which rays, if coming across a metallic object, such as another steamer or ship, would be reflected back to a receiver screened from the local transmitter on the sending ship and thereby immediately reveal the presence and bearing of other ships in fog or thick weather.'

Radar, the name by which this technology later became known, was, as Marconi saw it, potentially a useful tool for ships to locate other ships when visibility was poor. He also realised that it could be used to pick up the presence of metallic objects wherever they might be, including aeroplanes in the sky. In 1933, he gave a demonstration to the Italian military of how he could detect waves reflected by metallic objects. They listened but nothing came of it. Ugo Tiberio, a young Italian engineer and naval officer, who became a professor at the Italian Naval Academy under Mussolini, followed up Marconi's work. In 1935, Tiberio submitted a detailed set of technical proposals to the Italian military to solicit support for the development of two types of systems, what we would now call radar systems, that were based on continuous waves and wave impulses. Fortunately for the British and their Allies, the Italian military rejected Tiberio's proposal, as they were keener to put their money into developing another battle cruiser for the navy.

By contrast, when Sir Robert Watson-Watt made a similar technical proposal, the British military put their full weight behind his plans for developing a wave impulse system, which he called Radio Detection And Ranging (RADAR). During the Battle of Britain RADAR (later radar) meant that the RAF knew when the Luftwaffe was coming over the channel to attack in numbers, so defences were prepared before the Germans arrived.

The Royal Navy also used radar on its ships. It played a key role, for example, in Britain's victory over the Italians at the Battle of Capo Matapan in March 1941. This battle off the coast of Crete and southern Greece was historic as the British attacked at night, which had not been possible in naval warfare up to then, because radar allowed them to 'see' enemy ships. They had the advantage over the Italian navy, which was not expecting an attack overnight and had disarmed its main gun batteries.

The British battleships *Barham*, *Valiant* and *Warspite,* came to within 3500 metres of their enemy, completely unnoticed. The British got into position, lit up the night sky with flares and pummelled the Italians from close range. Inside three minutes two Italian heavy cruisers, the *Fiume* and the *Zara,* were destroyed and the *Vittorio Alfieri* and *Giosué Carducci* were sunk. After Capo Matapan the Italian navy belatedly decided to fund radar research.

Marconi joined the Fascist Party in 1923, the year after Benito Mussolini seized power after the march on Rome. Mussolini was the best man at Marconi's 1927 wedding to his second wife, Maria Cristiana Bezzi-Scali, the daughter of an Italian count. In 1930, Mussolini appointed Marconi President of the Royal Academy of Italy and he also became a member of the Fascist Grand Council. For the last decade of his life, from 1927 until his death in 1937, Marconi was an active fascist. In October 1935, Mussolini invaded Abyssinia and Italy became isolated politically from much of Europe. Marconi supported the invasion because he said it would bring civilisation to a backward people. He wanted to explain his views to the British people but the BBC refused to grant him airtime. Marconi responded that the BBC would not be broadcasting at all were it not for him.

Throughout the 1930s, his pro-fascist activities continued. When he died on 20 July 1937, after a heart attack, Benito Mussolini was at his bedside. The dictator gave Marconi a state funeral with full military honours in Rome and many thousands came out to line the streets and give him a final Fascist salute. He was buried on his family's estate.

There have been debates as to whether Marconi was a die-hard fascist, or whether, as a person of high status living under a dictator, he was forced to do what he did. He had voiced his concern for the wellbeing of mankind and believed radio had the power to bring people together but his support of fascism seemed to be at odds with this sentiment. A damaging report under the byline of Rory Carroll appeared in the *Guardian* newspaper on 19 March 2002, detailing Marconi's role in blocking Jews from becoming members of the Italian Academy. According to Carroll, documents had been found which showed that not one Jew was allowed join the Academy during Marconi's time as president despite the clear brilliance of some Jewish-Italian scientists. But it is not entirely clear whether Marconi initiated this policy or just went along with it.

There is no doubt, however, about Marconi's status as one of the great figures in the history of the human communications. His Irish roots and links to Ireland have been largely ignored, even in Ireland. For example, in 2009, Peter Kennedy, Professor of Electrical and Electronic Engineering in UCC, gave a public talk on Marconi and his influence on the hundredth anniversary of his Nobel Prize. He was disappointed when only a small number of people showed up. 'People don't associate Marconi with Ireland,' said Professor Kennedy. 'They think Marconi… Italian…and they switch off.'

The Legacy of Guglielmo Marconi

- He is considered the father of radio.
- He was the first to transmit a wireless signal across the Atlantic, from Cornwall to Canada.
- He was the first to build a commercial transatlantic wireless transmitter in Clifden.
- He was the first to define the different frequencies at which radio should be broadcast.
- He made a significant contribution to the development of radar systems.

11

PULSATING PULSARS

Jocelyn Bell Burnell (born 1943)

Birthplace: Belfast

It jumped up into her field of view and said, 'Yoo-hoo, here I am.' But what was it? The radio wave 'pulses' coming from some place in deep space were so unexpected that her first impulse was to check her equipment. There must be something wrong with the receivers. There wasn't. There must be some kind of interference causing the inexplicable readings, perhaps from a nearby radar station, communication satellites overhead or military aircraft operating in the vicinity? She ran some checks. Nothing. With the easy, explainable options exhausted, it was time to think the unthinkable. Could it be that this was an artificial signal? Was ET calling? Had an alien civilisation created the signal and beamed it towards earth, or at least in the direction of earth (let's not get carried away by our own importance here). If this was true, it would have been the biggest discovery in human history.

The other possibility, not remotely as dramatic but one that would be scientifically important, was that some unknown natural phenomenon was sending out these strong pulses of radio waves. It was August 1967 and whichever way it played out, life would never be the same for Jocelyn Bell, the twenty-four-year-old Northern Irish woman who had stumbled upon this scientific mystery.

As is often the case with game-changing scientific discoveries, Bell's discovery was 'totally accidental' and totally unexpected. 'I was doing the research project I had been set very conscientiously and happened across something unexpected,' she recalled later.

It was an exciting time for Jocelyn Bell when news of the dramatic discovery began to leak out from her University of Oxford lab. The press loved the story, particularly as there was talk of signals from space that couldn't be explained and the possibility that 'little green men' were trying to make contact with planet earth. These stories probably did not please many of Bell's conservative peers but she did not wilt under the strain. Indeed, she recalls quite enjoying the novelty of talking to journalists, whether for print, radio or television. She needed a way to describe her discovery in terms that ordinary people could understand. The story, as she tells it these days, goes like this:

'The analogy I use is imagine you are at some nice viewpoint making a video of the sunset and along comes another car and parks in the foreground and it's got its hazard warning lights, its blinkers on, and it spoils your video. Well my project was looking at quasars, which are some of the most distant things in the universe. They are big, big things like galaxies but they are incredibly bright and they send out a lot of radio waves, which is what I was picking up. [I was] studying these distant quasars and something in the foreground sort of went, "Yoo-hoo!" Not very loudly shall we say – it was a pretty faint signal – but it turned out after a lot of checking up and a lot of persistence to be an incredible kind of new star which we have called a pulsar, pulsating radio star.

'They are tiny as stars go, they are only about ten miles across but they weigh the same as a typical star so they are very, very compact.'

It is forty-six years since the discovery by Bell Burnell (as she has been known since her marriage in 1968) and scientists today know quite an amount about pulsars, although the precise way they work remains elusive. Pulsars are remarkable entities. They are about one and a half times the mass of our sun, so they have huge mass, but are so small they could fit into Galway Bay. Consider that planet earth could fit into the sun one million times. It is remarkable then that the sun, which is so large – about 930,000 miles in diameter – has less mass than a pulsar that could fit into Galway Bay, which is about 30 km wide by 50 km long. That gives an idea of the awesome density of a pulsar, which is about one million, billion times more dense than our earth.

Scientists believe that pulsars are formed during a supernova explosion at the end of a large star's life, a star at least eight times bigger than our sun. The light that is emitted during a supernova explosion, while a pulsar is being created, is so bright that it outshines the entire galaxy in which it lives. This is amazing considering that a galaxy can be enormous and contain up to 100 billion other stars, like our own sun, all with the potential to have their own solar systems. The phenomenal brightness of the supernova blowing up blinds the rest of the galaxy, until the explosion phase finishes and the star develops into a pulsar, or a rapidly rotating neutron star. A neutron star is defined as what's left after the gravitational collapse of a large star that essentially ran out of fuel. This process will not happen in our own solar system, as our sun is not big enough to result in a supernova explosion when it eventually runs out of fuel. Instead it will turn into something called a 'white dwarf', as it winds down towards its galactic demise. The death of our sun will be a violent event but not violent enough to produce a pulsar.

Essentially, neutron stars and their rapidly rotating cousins, the pulsars, are the remnants of the incredibly violent disintegration of a huge powerful star that ran out of fuel, heat and power. The aftermath of such colossal galactic destruction is what Bell was looking at when she saw the strong, regular pulses of radio waves appearing in her data set back in 1967. To use a non-scientific analogy, it was somewhat like being a witness to the devastation in New York in 2001 after the twin towers had collapsed. On that sunny September morning no one quite knew what was going on until the second plane crashed into the second tower. Was the first incident a one-off? Did a pilot have a heart-attack? Was it a suicide? Did some instruments fail? The suspicion that it was a terrorist attack was not confirmed until the second plane hit the other tower. After that, there could be no doubt: the US was under attack.

Similarly, when Bell picked up the now famous 'scruff' on her data set, she could not be sure that what she was seeing was not just a once-off. She needed to find another one or two to provide confirmation and ammunition against potential sceptics. The first pulsar was discovered on 28 November 1967, after Bell determined that the scruff on her data set was in fact a series of radio wave pulses, each separated from the next by about one second. The world's first pulsar was called CP1919 (don't scientists come up with the catchiest names?). Inside a few weeks Bell found three similar wave sources. She had found four pulsars, enough to prove that here was a real, new natural phenomenon. This also served to bury the notion beloved of the press that radio waves were being transmitted by little green men.

Bell Burnell recalls that she began to get excited when she found the second pulsar, not the first one. She felt after finding the second one that it looked likely that a new population of stars had been discovered and that these two might be just the tip of the iceberg. Her excitement was justified: 2267 pulsars have been discovered as of May 2013 and scientists expect to find thousands more.

As a young woman starting out a career in science in the 1960s, Jocelyn Bell was very far from the centre of power in the lab. The man calling the shots was Professor Antony (Tony) Hewish. It was routine then – and still occurs today – for the team leader to get the credit for anything that happened in his lab, while the PhD students did most of the grunt work.

Bell began her PhD in the University of Cambridge in 1965, under Hewish's supervision. The stated aim of her study was to study interplanetary radio waves and find out which sources were scintillating or transmitting strongly. These sources were likely to be coming from quasars, or quasi-stellar radio sources, to give them their full name. Quasars are among the brightest, most powerful and most energetic objects to be found in the entire universe. They are also very compact, moving fast and accelerating away from the earth. This distance indicates that they were formed in the early universe (as it has taken them a long time to get where they are) and are therefore very ancient. In the 1960s their exact nature was not known but since the 1980s the scientific consensus on quasars is that they are an incredibly bright cosmic entity that forms around black holes at the centre of certain galaxies and are 'fed' by material that falls into or is dragged into black holes. It is now thought that every galaxy has a quasar at its centre but not every quasar is bright and visible, as this depends on the nature of the material falling into its associated black hole.

The priority for Bell, when she started her PhD, was to identify quasars and to do this she searched for places in the sky that emitted strong scintillations of radio waves, which could be picked up by the big radio wave receivers that she helped to set up near Cambridge. A specially designed collection or array of receivers – looking like a group of huge modern satellite TV dishes – was constructed over two years in a four-acre field not far from the lab. Antony Hewish did the design work and the centre became known as Mullard Radio Astronomy Observatory. During the construction phase Bell spent most of her time getting her hands dirty, banging in stakes and stringing out and connecting reams of copper wire. When the array was ready for operation in July 1967, she began to monitor reams of paper-based data, looking for the tell-tale galactic signature of quasars.

To her great credit, Bell managed to pick out something important buried inside the ninety to one hundred feet of paper data she went through every night. She noticed something different in the data set of 6 August 1967. A paper in the respected science journal, *Nature*, eventually appeared, under the names 'Hewish, Bell et all', announcing the discovery of a new type of star, which the team called a pulsar. The paper was published on 24 February 1968 and earned Antony Hewish a share of the 1974 Nobel Prize for Physics, for what the Nobel Committee called 'his decisive role in the discovery of pulsars'. Hewish's co-winner was Martin Ryle, who was recognised for another aspect of the research.

Bell Burnell was not recognised by the Nobel Committee although she had been diligent in wading through mountains of boring paper data, alert enough to pick up something unusual and brave and determined enough to find an answer to the mystery of the 'scruff' when faced with the scepticism of senior colleagues. In the 1960s, to be a young woman in scientific research was not a powerful position. Times have changed and many believe that Bell Burnell's experience is part of the reason. These days science is better – although far from perfect – for young researchers and women.

Bell Burnell never expressed any sense of bitterness about the Nobel prize. She remained philosophical and took the path of trying to change things in science, particularly for women. This ability to roll with the punches and overcome setbacks was something she learned early on in life, when it was far from obvious that she would reach the heights she has attained in science.

Jocelyn Bell was born in Belfast in July 1943 but spent most of her first thirteen years in Lurgan. She failed the 11-plus, the examination that children in Northern Ireland took at the end of primary school (it was officially abolished in 2008). This exam was crucial as it usually determined whether a child was admitted to a grammar school where the focus was academic or a secondary school, where for many years pupils were not allowed to take A-levels, so could not qualify for university. Jocelyn Bell's failure in the 11-plus wasn't fatal for her future academic career, as she had already been attending the grammar school in Lurgan, which agreed to keep her on for a few years before she went off to a boarding school in England. She did admit much later that the failure 'shook her' and she didn't choose to mention it until she attained the status of professor.

Looking back today, Bell Burnell believes that the 11-plus curriculum didn't suit her, mainly, she says, because 'there wasn't any science in it'. Her scientific ability was certainly obvious when she came top of her class in her first term in secondary school at Lurgan Grammar. Before that, there was another hurdle to cross. Girls and boys were segregated in her first year of secondary school. At the time, Bell thought that the separation might have 'something to do with sport' but was horrified when she realised that the boys were being brought to the science lab, while the girls were being packed off to learn domestic science. In the 1950s girls in Lurgan and all over Ireland, south and north, were given little or no encouragement to do science. Bell's parents protested and she was permitted to join the boys doing science, along with the daughter of a local doctor and one other girl. It was a very close call and Ireland almost lost its most accomplished ever female scientist before she even had a chance to show what she could do.

Bell completed her two years in Lurgan Grammar, then it was off to England. Her family were Quakers and had a tradition of sending the children to Quaker schools in England, so she attended Mount School in York. She recalls that it was good to get away from home, although traumatic to begin with. In England, in the 1950s, there was a different attitude from what she had encountered in Ireland towards girls doing science. Bell did very well in her studies, despite what she recalls as a mixed standard of science teaching, and got a place in the University of Glasgow to study science. There she did well enough to be accepted to do a PhD in the University of Cambridge. It had taken a lot of determination for this quietly-spoken girl from County Armagh to make it this far.

After the paper in *Nature*, the phone hardly stopped ringing, with an endless round of interviews – which Bell Burnell describes as 'fun' – with radio, TV and print journalists. She had become an overnight scientific celebrity, while still only in her mid-twenties. It was difficult to settle back and finish her PhD but she completed it in September 1968. Bell Burnell said that one practical benefit of her new-found fame was that she never found it difficult to pick up a job when she was living in different places throughout Britain with her husband, Martin Burnell: he was a civil servant whose work took him from city to city. Burnell worked part-time for many years, following her husband's career, while raising their son Gavin (born 1973), who is also a physicist.

The downside of achieving success and fame at an early stage was that people expected her to come up with amazing discoveries all the time. A discovery such as pulsars comes only about once a decade among astronomers as a body, so it is a bit hard, she suggests, to live up to such expectations.

These days she continues to work as a visiting Professor of Astrophysics in Oxford University, where she is free to conduct research without too many other duties. Whatever she might do before she retires, her scientific legacy is secure. In 2010, a pulsar conference was held in Sardinia to honour Bell Burnell's forty-five years in science and 'christen' a new radio telescope. A long-time colleague, Australian pulsar researcher, Dick Manchester, was asked to deliver a speech at the conference outlining Bell Burnell's contribution to science.

He said: 'I think Burnell's fame is greater because she didn't receive the Nobel Prize in 1974 than it would have been if she had. I believe that the furore that her lack of recognition caused resulted in a change of attitude by the Nobel Committee and I'm sure more widely as well, with a heightened awareness of the role of students in projects and the role of women in science.'

The Legacy of Jocelyn Bell Burnell

- She discovered the pulsar, a new type of incredibly dense, pulsating star.
- Many scientists believe she should have shared the Nobel Prize for Physics in 1974.
- She is a role model for women in science, in Ireland and around the world.
- She helped to change attitudes about how the credit for research findings should be allocated.
- She is an advocate for the complementary role of science and religion as, to her mind, both seek a greater understanding rather than seeking truth.

12

SPIRAL GALAXIES

William Parsons, Third Earl of Rosse (1800-67)

Birthplace: York; home in Parsonstown (now Birr), County Offaly

It is 1810 and a stagecoach with two passengers bounces into the Berkshire town of Slough, twenty-two miles from central London. William Parsons, a ten-year-old Irish boy, is on the last leg of a long and arduous coach and boat trip from his home in Parsonstown to the capital city with his father, the Second Earl of Rosse. The Earl travels regularly to London, where he has a seat in the House of Lords. The boy has made the trip with his father many times and, for him, the most exciting part of the entire journey is just about to happen. Slough is one of the appointed stops where the stagecoach horses are changed on the journey between Bristol, where the boat from Dublin puts in, and London. The two travellers, as by now is their wont, decide to take the opportunity provided by the change of horses to pay a visit to the town's biggest tourist attraction: the telescope designed by German-born English astronomer, Sir William Herschel. The so-called 'Great Forty-Foot', built between 1785 and 1789, was the world's largest telescope for almost fifty years and Herschel became famous for his discovery of the planet Uranus and two moons of Saturn.

Young William is awestruck by the sight of the telescope, although he has seen it many times. He looks it up and down before walking slowly to the viewing area. There he peers down its great lenses and we can imagine him thinking, 'One day I will build a bigger telescope than this and discover new things beyond our own solar system.' This boy will become Ireland's most famous astronomer. In 1845, he will build the greatest telescope the world has ever seen, which will become known as the 'Leviathan' (a biblical monster). He will use it to make the historic discovery that our universe is made up of more than one galaxy and that galaxies can have a spiral, whirlpool structure.

In 1800, the year William Parsons, the Third Earl of Rosse, was born, mankind had only begun to understand the solar system in which we live, let alone what exists outside of it. Scientists had managed to measure how far earth was from the sun and had a good idea how big our solar system was. But we had no idea how far it was to the nearest star, nor did we have any idea of the size or the age of the universe. From an astronomical point of view, mankind was still in the Dark Ages.

Sir William Herschel, perhaps the most famous astronomer of the age, was coming to the end of a twenty-year effort to catalogue thousands of distant objects known as nebula. These had been known about for hundreds, even thousands of years: Ptolemy, the Egyptian-Roman mathematician and astronomer, had described seeing something he called nebula as far back as 150AD. In 1800, astronomers still didn't know what these nebulous celestial features composed of reddish gas were all about. Given the limitation of telescope technology at the time, astronomers like Herschel had the task of trying to focus on something that remained blurred and elusive – nebulous.

Herschel worked tirelessly with the French astronomer, Charles Messier, and discovered about 2400 objects in the cosmos which were deemed to be nebulae. They were catalogued as part of the first ever deep sky survey. It was a heroic effort but when Herschel died in 1822, he was no closer to getting a clearer astronomical image of these mysterious fuzzy objects or to understanding scientifically what they were.

The definite solution to the mystery would not arrive until more than a hundred years after Herschel's death. In 1924 the American astronomer, Edwin Hubble, was in a position to confirm that nebula were, in fact, galaxies. This was a profound moment for science: the first proof that galaxies existed beyond our own – lots and lots of them. We were no longer alone. The Milky Way had millions, if not billions, of celestial brothers and sisters.

In this epic story of human discovery, from Herschel to Hubble, William Parsons made outstanding contributions. In 1845, as the Great Famine was ravaging Ireland, he unveiled his monstrous Leviathan and began to point its huge lenses at the sky from its spot in the grounds of the Parsons family's ancestral home.

The Leviathan had overtaken the Herschel telescope in Slough as the largest in the world. It was a monumental feat of engineering. The bronze reflecting mirrors it held were a world-record 72 inches in diameter, the viewing tube was 54 feet long and weighed 12 tons. The question was; would the Leviathan be any better at viewing nebula than Herschel's telescope was?

This story begins with a nebula, somewhat unfortunately named Messier 51 (M51) which was discovered by the afore-mentioned Charles Messier in 1773. M51 is located about twenty-three million light years from our Milky Way in the constellation of stars called Canes Venatici. The interstellar distance from the Milky Way to M51 is mind-blowing, considering that a light year is the distance that light can travel in one year and that light moves at about 300,000 km per second.

However, in terms of the vastness of space, M51 is our galactic neighbour. Parsons knew this and M51 became a target for discovery. It wasn't too distant and was relatively easy to pick out in the night sky, so he decided that he would build his telescope specifically to get a better view of it. There was nothing accidental, therefore, in what was to become his great discovery.

Soon after the Leviathan became operational, Parsons focused its giant lens on the area of sky that he knew was home to M51. He had no idea whether his telescope would do a better job of seeing this nebula than Herschel's had. He didn't have to wait long for an answer as features of M51 came into view, although he had to strain to make them out. He grabbed a pencil and began to sketch. The drawing he produced of M51 – which today is more commonly known as the whirlpool galaxy – remains strikingly accurate, even compared with images taken by the vastly more powerful Hubble telescope. With his keen eyesight Parsons picked out spiral features and something that looked like a whirlpool. He was the first human being to see the swirling features we now associate with galaxies.

Most people peering at M51 through a telescope with technology like the Leviathan would distinguish no more than a small dot or blotch of light. Parsons accurately picked out the details and he was proved right. The sketch of M51 made him and his huge telescope famous worldwide.

Parsons was somewhat perplexed by the spirals and whirlpool features he was looking at, however, and thought they might be representing clumps of stars that had joined together. It was not until Hubble linked nebula and galaxies that the significance of his findings would become apparent.

The ancestors of William Parsons had arrived in Ireland from England in the 16th century and secured substantial land holdings around what is now Birr. They were progressive landlords and generally well liked. Members of the family showed a talent for science and engineering down the centuries and the first generation of Parsons in Ireland were cousins to Robert Boyle (*see* Chapter 13), considered the father of chemistry. Two individuals by the name of Parsons feature in this book: William in this chapter and his son, Charles (*see* Chapter 6), who invented the steam turbine engine. Another of Parsons's sons, Richard Clere, was an engineer who built many of the first railways in South America as well as the Buenos Aires sewage works, which helped to rid the city of regular outbreaks of deadly yellow fever.

This interest in and aptitude for technical subjects and science were highly unusual among landed gentry in Ireland and Britain in the period from the 16th to the 19th centuries. It was more the norm for the younger sons of big landowners to pursue careers in the law, the clergy or the military.

Parsons's father, Laurence, married Alice Lloyd in 1797. They had five children, Jane, Alicia, William, John and Laurence. William was born on 17 June 1800. At this time, memories of the French Revolution were still vivid in the minds of the British establishment, who feared a similar occurrence in England. The Rebellion of 1798 intensified these fears. The Parsons children, who were educated at home, were taught French, a political statement that indicated a free-thinking, libertarian cast of mind on the part of their father. Laurence Parsons was a friend of Wolfe Tone, a leader of the 1798 Rebellion. In 1807 he inherited a peerage from an uncle in New Ross, head of another wing of the family, and became the Second Earl of Rosse, with a seat in the House of Lords.

The Second Earl had demonstrated his own interest in engineering when he visited the German town of Baden-Baden to see the world's first metal suspension bridge, which had been built there in 1799. In 1826, he built his own suspension bridge over the Camcor River in his demesne in Parsonstown. This is turn inspired the designers of the Brooklyn Bridge, linking Manhattan and Brooklyn, which became the longest suspension bridge in the world when it opened in 1903.

The Second Earl left his children a vital legacy: he taught them the techniques required to work in the estate's forge. In the early 19th century every decent-sized estate had a forge where blacksmiths shod horses, repaired wheels and metal implements and kept various carriages and carts in working order.

The knowledge Parsons gained in the forge contributed greatly to his scientific success. He learned how to mix materials, how to make things, how to adjust mixes to enhance particular properties in a product. These skills, combined with his great natural scientific ability, later enabled him to build the largest telescope the world had ever seen.

Parsons was fortunate to have been born into a family that gave him the means to achieve his dreams. He was something of a late starter and his major achievements didn't come until he was into his mid-forties. He spent his youth on his family's magnificent 1200-acre estate, where he and his siblings were educated by their father and by private tutors. He went to TCD where he excelled in mathematics, winning an academic prize. As an undergraduate he transferred to Magdalen College, Oxford, where he graduated with a first-class honours degree in mathematics in 1822. Despite his undoubted ability, he does not appear to have achieved much of note during his twenties.

A chance meeting with a like-minded person at the Royal Society in London changed everything. In the early 19th century the Society provided public lectures on all aspects of science and engineering and its lecture rooms became a fashionable place for well-heeled people interested in science to meet. Here, Parsons met Mary Field, an immensely wealthy Yorkshire heiress thirteen years his junior, whose family had vast land holdings going back to the Middle Ages. Mary was also an extraordinary character in her own right, a talented pioneering photographer who was interested in science. It was another stroke of luck for Parsons.

The couple married in 1836 and had four sons: Laurence and Randal, as well as the aforementioned Richard Clere and Charles. Parsons became an enormously wealthy man overnight when he assumed control of his wife's estate, which had revenue equivalent to about €10 million per year today. His wife, in return, gained a title, Countess of Rosse. There may be a temptation to dismiss the marriage as one of convenience but it does appear that the couple were perfectly matched, with similar interests. Now aged thirty-six, Parsons no longer had to worry about money. He could devote his working life entirely to his passion for astronomy and science. Mary Parsons became a fellow-traveller and she was often by his side in the forge, where she showed remarkable skills as a blacksmith, something unheard of for a woman of her class, or any class for that matter, in the mid-19th century. She was responsible for much of the iron work that went into the Leviathan. She played a major role in her husband's success and he seemed to be spurred on to much greater achievements after his marriage.

Parsons became the Third Earl of Rosse in 1841 on the death of his father. With his wife's financial might, he was in a position to think about how to build a better, bigger telescope than Herschel's – something he had dreamed of doing since boyhood. This was a challenge, as Herschel's reflector telescope was one of the wonders of the scientific world, although completed in 1789.

The challenge Parsons faced was – put simply – to build bigger reflector mirrors than Herschel had managed to do, something that wouldn't be easy. Reflector telescopes work by attracting light down a tube to where it hits a slightly curved mirror, which reflects or bounces the light towards a secondary mirror. The light bounced off the secondary mirror passes through various lenses and onto the human eye. The basic rule is that the more light that can be brought in, harnessed and reflected, the further and clearer the telescope can see.

Herschel built a 49½-inch reflecting telescope for a very good reason as, in 1789, instrument makers couldn't make bronze reflecting mirrors any bigger than this. When they did try to build them, by melting bronze in an oven, pouring the liquid into a mould and waiting for it to solidify, the mirrors cracked. There was a natural limit, it seemed, to the size of mirrors made from bronze.

It could be argued that Parsons's true genius was his ability to work with materials, something he had learned by working in his father's forge rather than in TCD or Oxford. He believed he knew how to make larger bronze mirrors that wouldn't disintegrate on cooling. If he could do this, the way was open for him to build the world's largest telescope.

His ingenious solution to the problem that affected bronze was to find a way to slow down and control the cooling process, making use of two ovens, not just one. This use of a secondary, so-called 'annealing' oven, provided him with far greater control over the speed at which liquid bronze cooled, something that was crucial. The first part of the process stayed the same – 4.5 tonnes of bronze metal melted in an oven – but here the similarities with Herschel's method ended. Herschel simply poured the molten metal into a mould of the size he wanted. Parsons made the technical improvement of adding a grid inside the mould, so that blocks of metal, once they solidified, could be easily moulded together. At a certain temperature, the blocks of solid bronze separated by a grid inside the mould could be fused.

As well as the grid inside the mould, the annealing oven itself was a major advance. It kept the bronze molten and allowed for a slower, controlled solidification. The net effect of these innovations was that the cooling process changed. Inside of cooling starting from the outside, which caused cracking, it started from inside the core. The bronze slowly cooled from inside out, as it was poured into the mould grid, eventually allowing a bronze mirror larger than ever before to be constructed.

This was a major breakthrough in the field of what would

nowadays be called materials science, in which Ireland still excels. In 2010, the country was ranked eighth in the world in this area.

It took Parsons several attempts using the new methods to make bronze mirrors. The first couple of mirrors cracked and disintegrated on cooling but Parsons persisted. His first achievement, in 1840, was to use his new process to build a scaled down version of Herschel's 49½-inch telescope with a 36-inch bronze mirror. That was an important first step but it would take Parsons five years to get to the next level by building a mirror bigger than Herschel's.

This construction of the Leviathan took more than three years and cost £12,000 (about €2.4 million today). Parsons's goal was to make a mirror that was 72 inches in diameter. As well as the sheer size, there was another technical challenge to be addressed. How would he know that the huge mirror of his new telescope was properly polished? If it wasn't polished properly it would not be able to 'see' properly. The focal length of the mirror was 65 feet: that is the extent of the area where light entering the telescope is gathered and reflected. The gathered light is reflected onto a particular spot. The more light that is gathered and the greater the focal length, the further the telescope can 'see'. He needed a structure that was 65 feet high to test whether the telescope was working.

The tallest structure on the Parsonses' estate was the tower, 45 feet tall from its summit to the bottom of an open trapdoor where the newly manufactured bronze reflector mirror could sit. But the focal length of the mirror was 65 feet so Parsons was 20 feet short. The solution he devised was to build a 20-foot mast that sat on top of the tower, extending it to the required height. At the top of the mast, he placed his pocket watch.

This was an ingenious idea: he knew that if he could read the time of his watch, the mirror was working. The light would enter and be reflected onto a focal point 65 feet away, which was the

72-inch bronze reflector mirror. This would bounce light towards a second mirror, which, in turn, would reflect light towards a viewfinder. Parsons would view the pocket watch through the viewfinder at the top of the tower, through a series of corrective lenses. When he did this, he could read the time on his watch.

Like many scientists, Parsons was a perfectionist and realised that the focal point – the reflector mirror – could vary because of variations in temperature. For example, if it was a particularly hot day the mirrors could expand slightly. He decided to put a specific adjustment on the telescope to take care of this variable.

Then there was the problem of the weight of metal being used. Parsons set out to make a mirror as large as possible. However, the biggest he could achieve was 72 inches as, after that, the sheer weight of the metal would distort the mirror and blur what could be seen in it.

There is no doubt, despite his liberal views and support for Catholic Emancipation and reform, that the Third Earl lived in a different world from the millions of impoverished Catholics in the country. In 1845, while Parsons was focused on the completion of his telescope project, the potato crop failed and people began to starve in their hundreds of thousands. Although building his giant telescope had been his dream for a long time, he put this work on hold for a year after its completion, instead devoting all his scientific energies and abilities to trying to find the cause of potato blight and, he hoped, a remedy. Remarkably, he came close to finding a cure, discovering that a substance called copper sulphate was capable of killing the fungus that was causing the blight. This was promising but there was a complication that he could not resolve: copper sulphate was also poisonous to the healthy potato plant.

Parsons also employed as many local people as possible during the years of the Great Famine, some six hundred people at the peak. He set up a soup kitchen on his estate and organised the construction of a new building, where Birr Castle Museum is located today, as another means of famine relief.

In the decades after its construction people came from all over the world to visit the Leviathan. While Parsons was alive, it was no doubt a thrill to be shown the Leviathan by the man who created it. It remained the largest telescope in the world until 1917 – seventy-two years – but in reality its best days were long behind it, long before it lost its title as the world's biggest.

After the death of Parsons's son, Laurence, the Fourth Earl of Rosse, in 1908, the now neglected Leviathan was partly dismantled. In 1914, one of its mirrors was moved to the Science Museum in London, while the other mirror vanished. For the next eight decades, the remains of what was once the Great Telescope lay forgotten in Birr. It had become scientifically obsolete in an age when photographs became a more and more important part of astronomy. The Leviathan was limited as it could focus on only one part of the sky at any one time and required two minutes of continuous focus on that single spot to generate a photograph. New telescopes appeared that could take photographs continually of all areas and track the sky by moving in a perfect arc from east to west. The Leviathan's days as a working research telescope were over and the story of the Third Earl and his telescope forgotten by most people.

In 1996, however, renewed interest in the telescope led to restoration work, which involved recreating the original structures of the telescope as closely as possible. These days the Leviathan again stands tall and looks to the heavens in Birr.

The Legacy of William Parsons

- He completed the world's largest telescope, the Leviathan, in 1845
- His telescope was the world's largest for seventy-two years.
- He was the first to see a galaxy outside our own Milky Way.

The statue of William Thomson, Lord Kelvin (1824-1907) in Kelvingrove Park, underneath the tower of University of Glasgow. (Photo courtesy of Matthew Trainer, University of Glasgow)

The yacht Turbinia, *which was powered by a steam-turbine engine patented by inventor and engineer Charles Parsons (1854-1931).*
(Photo courtesy of Tyne and Wear Archives and Museums/The Bridgeman Art Library)

Italian born of an Irish mother, Guglielmo Marconi (1874-1937) with his wireless radio machine.

The remains of Marconi's wireless transmitting station in Clifden, County Galway, which began operating in 1907 and employed two hundred people at its peak.
(Photo courtesy of Peter Kennedy,
Professor of Electrical and Electronic Engineering, UCC)

The Titanic *in Cobh on its maiden voyage in 1912. The US Radio Act of 1912 stipulated that all passengers ships should have round-the-clock wireless radio communications.*
(Photo courtesy of Cobh Heritage Centre)

Ernest Walton (1903-95), the physicist who split the atom and Ireland's only Nobel laureate in science. (Photo courtesy of Philip Walton)

Vienna-born but an Irish citizen, Erwin Schrödinger (1887-1961) lecturing in the Dublin Institute for Advance Studies. (Photo by the Irish Press*, courtesy of the DIAS)*

William Wilkins, headmaster of the High School and grandfather of Maurice Wilkins (1916-2004), Nobel prizewinner and DNA's 'third man'. (Photo courtesy of the High School, Dublin)

Jocelyn Bell Burnell (born 1943) who discovered pulsars (pulsating radio stars) in her Cambridge laboratory as a young PhD student in 1967.
(Photo courtesy of Jocelyn Bell Burnell)

- He was the first to detail the spiral nature of galaxies.
- He was the first to gain an insight into the structure of nebulae.
- He discovered the whirlpool galaxy, also known as M51.
- He greatly advanced telescope-building technology.

PART IV

EXPERIMENTS, EVOLUTION, LIFE AND LOGIC

13

THE GREAT EXPERIMENTER

Robert Boyle (1627-91)

Birthplace: Lismore, County Waterford

In the ancient world, the general understanding of nature and how the world worked came from ideas formulated in the minds of great men, ideas that were not subject to rigorous test or challenge. The Greek philosopher, Aristotle, devised a complicated theoretical view of the world, based on the interactions of what he called the four elements: fire, air, water and wind. Aristotelian notions of nature held sway in the west for the best part of two millennia – remarkable given there was no evidence to support what Aristotle was claiming. For centuries, a largely uneducated populace accepted either the truth as set out by the ancient philosophers or by their Church leaders. This was truth inherited through tradition.

In this world, theories of how things worked were not subject to the scrutiny they come under today. In fact, scientists did not routinely begin to test new ideas through the use of experiment until the latter part of the 17th century. There were many great figures of science, from Copernicus to Francis Bacon to Descartes, who helped put experiment at the core of all science. However, the person who did more than any other to achieve this was Lismore-born Robert Boyle, seventh son and 15th child of the ruthless Tudor adventurer, Richard Boyle, the first Earl of Cork.

The edifice of ignorance about how the world worked, which began with Aristotle in the 3rd century BC, was very much still in place as the 16th century began. The first ray of light, pointing towards a new age of enlightenment, came in 1543, when the Polish astronomer, Nicolaus Copernicus, claimed in his book *De Revolutionibus Orbium Coelestium* (*On the Revolutions of Celestial Spheres*) that, based on his observations and calculations, he believed the earth revolved around the sun, not vice versa. This was a key moment in the history of science and marked the start of what became known as the scientific revolution. The Catholic Church saw Copernicus's claims as contradictory to the Bible's view of the earth being at the centre of the universe and the power of the Church meant that supporting Copernicus was a dangerous business. Nevertheless he *was* supported by other men of science – famously Galileo Galilei, the Pisan astronomer, who, in 1633, was brought before the Roman Inquisition, convicted of heresy and placed under house arrest for the rest of his life.

The Copernican controversy rumbled on for decades and became the touchstone for a Church-science 'world view' struggle that arguably continues in other forms right up to the present. The 17th century French mathematician and philosopher, René Descartes, who also supported Copernicus, declared: 'I think, therefore, I am.' This was his way of saying, 'The evidence for my existence is not based on a mere belief but on the fact that I am capable of thought.' Then in 1620, Francis Bacon, the English scientist and philosopher, also in the Copernican camp, published his *Novum Organum* (*New Instrument*). This was another important moment in the history of science, as it helped push the case for experimentation as a core part of science. In this book, Bacon set out the first clear set of rules for how a new science should be conducted, based on three steps: observation and data gathering, formulation of theory based on that data; and testing of that theory through experimentation. Bacon's book had a massive influence on Isaac Newton and Robert Boyle.

These days Boyle is remembered almost exclusively for 'Boyle's Law'. This is unfortunate, as it distracts from his real genius and it was not his law to begin with – but more of that later. His importance lies not in that single, experimental proof of an idea formulated by others but in how he established widespread use of experiment in all sciences through his work in the laboratory, his publications (some ground-breaking), his fame and his co-founding of the Royal Society in London in the 1660s. This society was set up to promote and support excellence in science and to put experiment at the heart of it all. It was very influential, as it was set up by many of the great minds of British science at the time, and the public paid attention to what it did and what it said. Boyle was a key figure driving this. His reputation as a scientist and his meticulous work in the laboratory were widely respected.

This contribution would have been impressive enough for one individual but Boyle will also be remembered for establishing the 'mechanistic' view of the world. He saw the world as being like a giant machine, with lots of moving parts, big, small and very small, all interacting and subject to the same simple physical laws. The secret to understanding the world, as far as Boyle was concerned, was to learn more about the parts and how they worked together. This approach – understanding structure as a way to understand function – became central to science as time went on. For example, three centuries years later, when James Watson, Francis Crick and Ireland's Maurice Wilkins (*see* Chapter 16) were seeking to unlock the secrets of DNA, they focused on determining its structure. They believed that if they knew the precise structure of DNA they could determine its function, and that is what the three scientists managed to do.

Boyle's Law, which describes the relationship between

pressure and volume on a gas, is famous today because it appears in school textbooks the world over but in the broader scheme of things, it is just one of the many laws describing how gases behave. In other words, it's not all that important. What we know as Boyle's Law, should, strictly speaking be called Towneley and Power's Law, after the two amateur English scientists who formulated it in 1661. The idea for the law was born when these two gentlemen were on a study trip out in the Lancashire countryside, with scientific equipment and samples in tow, including a trapped air bubble. They walked up a hill called Pendle Hill and noticed that the trapped air bubble increased in volume as they climbed. The two men thought about this and put forward the hypothesis or theory to explain their observations that later became Boyle's Law.

The men proposed that the volume of a gas (or air bubble in this case) would increase if the pressure on it decreased and the opposite would also hold true. So, as the men walked up the hill, the atmospheric pressure on the air bubble decreased, as there was less atmosphere weighing on them as they climbed. This afforded the bubble more freedom to expand in volume. Boyle heard about what he called 'Mr Towneley's hypothesis' and felt it was interesting enough to test in his laboratory. Here Boyle proved the hypothesis correct but he was fair-minded enough not to claim the law as his own.

The Boyle family story in Ireland goes back to the time of the Tudor conquest, the colonisation or 'plantation' of the country, which was carried out mainly in the reigns of Mary Tudor and Elizabeth I. In the course of the 16th and early 17th centuries, much of the land of Munster and Connacht passed from the native Irish to English 'adventurers' or soldiers, rewarded after successful military campaigns.

Boyle's father, Richard, the 'Great Earl of Cork', had left England as an adventurer in 1588, aged just twenty-two. At the time there were rich pickings for ambitious men like Richard, as the Crown went about the systematic subjugation of Gaelic Ireland. It is said Richard arrived in Ireland with just £27 in his pocket, several suits, a sword and a dagger. He proved to be an excellent, ruthless businessman. He made connections quickly and was appointed Clerk of the Council of Munster by Elizabeth in 1600. This provided him with an opportunity to buy the large estates of Sir Walter Raleigh in Cork, Waterford and Tipperary, which had been acquired through the dispossession of original land owners, after Raleigh fell from favour with the queen. He continued to buy land opportunistically at knock-down prices as conflict rumbled on in Ireland into the reign of the first Stuart king, James I, which began in 1603.

The Earl of Cork was in his sixties and his wife, Catherine Fenton, in her forties when their son Robert was born. The Earl had spent years accumulating wealth and power and was determined, in his declining years, to consolidate these gains. His main way of doing this was by arranging strategic marriages for his sons and daughters with members of other powerful, wealthy families. Boyle saw his siblings married off, one by one, in this fashion. However, when his turn came and the Earl tried to arrange his marriage to Anne Howard, enticing her with gifts and jewellery, Boyle resisted. They were both still in their teens but Boyle defied his father and the Earl died before he managed to force his son's hand. This unpleasant experience was perhaps one of the reasons Boyle never married.

The evidence gleaned from Boyle's extensive surviving correspondence (letters to friends, family and colleagues) suggests that he was uneasy about how his father had accumulated his vast wealth, fearing that the money that allowed him freedom to pursue science, had not come from entirely licit sources. In fact, it would have been difficult for Boyle to avoid such a conclusion, given that his father had been imprisoned at one stage on charges of embezzlement and later subjected to heavy fines for possessing defective land titles.

Boyle was raised by a wet nurse, as was the tradition in wealthy families in England and Ireland in the 17th century. At first he was educated at home by a tutor. He had an odd way of talking and stammered to such an extent that some found it amusing. He was an austere figure, highly religious, and spent a large part of his life living with an older sister, Katherine, Lady Ranelagh.

When Boyle was eight, he and his brother Francis were sent to school to Eton, which was becoming a fashionable place for wealthy families to send their sons. The headmaster at the time was John Harrison and the importance of the Earl is shown by the fact that the brothers lived in the headmaster's house while they were at Eton. Boyle's time at Eton went smoothly until Harrison died. The new headmaster was strict, it seems, and Boyle did not flourish under the new regime so their father decided to remove the boys from the school when Boyle was eleven. He went back to live in Lismore and again had a private tutor.

It might seem incredible to us now that the Earl sent Boyle at the age of twelve and his brother, only a few years older, to do a grand tour of the Continent. The boys did have their tutor in tow, to watch over them and guide them. The idea of a grand tour for wealthy young Englishmen became popular later on, in the 18th century, so the Earl of Cork was ahead of his time as well as being very wealthy. The general idea of the grand tour was to encourage young men to learn more about foreign lands as well as science and the arts generally. For Boyle, the tour was about education in the widest possible sense.

The Boyle boys visited Dieppe, then Paris and Lyon, before arriving in Geneva. After that, the next stop was Italy and Boyle learned Italian before visiting Venice and Florence. The astronomer Galileo died in 1642 while Boyle was in Florence.

It was a major public event and left an impression on him. He later decided to study the work of Galileo, who influenced him strongly. In particular, Boyle would have admired how Galileo resisted pressure from the Catholic Church, when he declared that the sun, not the earth, was at the centre of the known universe. From then on, Boyle developed a strong belief that the world could best be understood through mathematics and mechanics and that declarations of faith, made without evidence, were completely pointless.

The boys came back to Marseilles. From there, in the middle of 1642, they planned to return home. The problem was, however, that a rebellion against the Crown had broken out in Ireland in late 1641, initially in Ulster, which quickly led on to a nasty conflict between native Irish Catholics and English and Scottish Protestant settlers. The boys' father fought the rebels and was reportedly disgusted when King Charles I, who had his own problems with parliament at home in the lead-up to the Civil War, negotiated a peace with them. This, to the Earl's mind, was treating the Irish with a respect they didn't deserve.

The Earl of Cork died in 1643. The boys had been away from home for four years and were now living in Geneva, still with their tutor. Boyle, aged seventeen, returned (to England) in 1644, after selling some jewellery to finance his trip home. He inherited a house in the Dorset village of Stalbridge on his father's death. There he planned to start pursuing the scientific interests that had been awakened on his grand tour. Boyle's first problem was that England was in a state of chaos. By now civil war was raging between the King's forces and those of parliament and travelling the countryside was difficult and dangerous. It is just over a hundred miles from London to Stalbridge but it took Boyle four months to get there, simply to claim his estate. He didn't finally settle in Dorset for another two years, during which time he lived in London with his sister Katherine.

An interesting aspect of Boyle's life and career is his strong religious beliefs and how he reconciled them with science. This story begins in Geneva, while young Boyle was still on his grand tour. Apparently there was a huge storm, which terrified him. When the storm abated and he realised that he had survived he began to wonder why he had not been struck by lightning. He concluded that the Almighty had preserved him for a special reason, to reveal God's work through experiments. For Boyle, a God that created a mechanical universe, which worked like a giant machine, with its parts in constant motion and obeying simple mathematical laws, was more to be admired than a God who would create a disorderly universe.

Boyle was a scrupulous man and stuck to his principles even when doing so was against his interests. For example, in 1680 he declined the offer to serve as the President of the Royal Society, although he was a founding member, along with a number of his close colleagues. His reason was that he could not swear the oaths of loyalty that were required in the reign of Charles II.

Boyle believed that his scientific investigations could provide him with spiritual insights into the nature of God and his work but he was worried by the possibilities that alchemy presented. He believed in alchemy, the 'science' that held that base lead or zinc could be converted into gold. He was convinced that it was possible to do this but worried that manipulating materials in this way was against morality and represented temptation by the devil.

Despite Boyle's strong religious faith, like many people, he had doubts. He was a worrier, it seems, concerned about things like whether he had treated his servants properly, given enough money to charity, or even whether he had benefitted wrongly from his father's ill-gotten gains. He was incredibly exact and precise in his experiments, many of which still stand the test of time, and he may have applied the same forensic level of examination to his moral and spiritual life.

Boyle began intensive studies when he started to live permanently in Stalbridge from 1646. The work kept him busy but he was lonely, which was perhaps what prompted him to make frequent trips to London, to see his sister Katherine and meet people who were embracing the 'new philosophy'. This philosophy centred on putting ideas to the test, no matter to which field they belonged, to establish whether they stood up to scrutiny. If they didn't, they should be discarded or reformulated. There was an influential emerging group, based in London and Oxford, which Boyle called the 'Invisible College' that adhered to these principles. The members of this college, including Boyle, would become the Royal Society of London, the first scientific society of its kind in the world.

Boyle was apolitical and had no real preference for one side or other in the Civil War. His father was a committed Royalist, whilst Katherine supported the Parliamentarians. He was lucky with how it all ended. Charles I was defeated and executed, Cromwell emerged the undisputed victor and was determined to exact retribution on his enemies, including Irish royalists. This meant that even more land was confiscated from the native Irish and passed to English colonists. Boyle was a huge beneficiary of this land redistribution and became vastly wealthy from 1652 onwards.

He spent two years in Ireland, 1652-54, organising his estate, which meant consolidating the holdings gained following the Civil War. Aged just twenty-seven, he returned to England, financially secure for life. The huge earnings from his Irish estate would support his lifestyle in London from then on and give him the opportunity to devote his time and energies to the areas of most interest to him. In defence of Boyle it should be said that he was reportedly generous with money to those around him.

Boyle would probably have considered his nationality to be English and was referred to in his later years (which were spent almost exclusively in London) as the 'English philosopher'. There have been suggestions from some quarters that he was a bigot but his later correspondence shows that he had many friends from across the Judeo-Christian spectrum.

As time went on, Boyle spent less time in Stalbridge and more in London with Katherine and in the company of his fellow 'Invisibles'. There was a certain amount of pressure on scientists of his kind during Cromwell's rule (1649-58). The Puritan leadership was not keen on men of science questioning and experimenting with nature, God's creation. John Wilkins, one of the leading Invisibles, was the warden of Wadham College, Oxford, and a supporter of Cromwell. He encouraged Boyle to move from London to Oxford, where he would enjoy more freedom to practice his new experimental science. Wilkins was strongly in favour of tolerance and brought many talented people to Oxford, particularly those interested in experimental science. Boyle lived in Oxford from 1654. There he began conducting experiments and benefitting from encountering such great minds as Christopher Wren, the architect of St Paul's Cathedral, and the mathematician John Wallis, chief code-maker or cryptographer for the English parliament and later for the royal court.

Oxford was a hugely stimulating environment for Boyle but London continued to also be a draw, particularly after the Royal Society was founded there in 1660. He moved back to the city in 1688 to live with his sister in Pall Mall. He was forty-one and would spent the remaining twenty-three years of his life in London. Boyle was sufficiently wealthy to employ Robert Hooke, a famous scientist in his own right, as his research assistant and Hooke built a laboratory for him in the garden of his sister's house. In 1665 Hooke invented and described the first microscope that was used to look at biological materials.

Hooke and Boyle worked very well together as experimentalists in their garden laboratory. One thing that fascinated them both was the air that we all live in, as Boyle put it. They set up a range of experiments, using an air pump, to test and define some of the properties and characteristics of air. A number of classic experiments followed. They showed, for example, that sound was carried through the air, that a candle needed air to continue burning and that liquid mercury in a tube could be drawn upwards by air. The latter proved to be a very good way of measuring air pressure and we still hear the expression 'the mercury is rising' when the weather is improving. They also created experiments for testing for the existence of a vacuum. A vacuum, in scientific terms, is space empty of matter. There was a debate at the time as to whether a vacuum could exist or not and many scientists believed it was not possible. Boyle believed a vacuum could exist, a belief based on his own observations.

Boyle's most famous book – and he wrote many – was *The Skeptical Chemist* (1661). This book is one of the main reasons he has been referred to as 'The Father of Modern Chemistry' since the 19th century. In it, he argued against the Aristotelian view that all chemical elements should be placed in one of four classic categories: earth, fire, air or water. He put forward the idea that matter was composed of tiny particles, or 'corpuscles' – clusters of particles. Boyle argued, like Descartes, that the world was a complex system made up of groups of small particles, governed by simple mathematical laws. In order to understand the world, it would be necessary to find out what the particles were, how they fitted together and the rules that governed their existence.

He also argued for the rigorous use of experiment to test theories in science and for chemistry to come out of the historical shadow of medicine and alchemy and become a new science in its own right. Up to then, chemistry had been considered a lowly, dirty activity. The true importance of chemistry to Boyle was that he could design chemical experiments to prove his 'mechanistic' view of how the world worked.

He could use chemistry to show that stuff, or matter, was indeed made up of small particles and that substances could be changed from one form into another. For example, he did some interesting experiments on saltpetre, or nitre, which is found naturally in very dry areas like the Atacama desert of Chile and is used today as a fertiliser (it was historically used as gunpowder). Boyle showed that it was possible to separate the component parts of nitre and recombine them. He believed that this showed how natural parts could be broken up and recombined.

Boyle is credited with being the first to use the term 'chemical analysis' in the way that it is understood by chemists today. At the time he wrote *The Skeptical Chemist* there were no university chemistry departments in existence. He succeeded in raising the status of this field of study, as everyone respected his meticulous and accurate work.

He published an analysis of mineral waters considered to be one of the best ever written on the subject, as well as an analysis of human blood. His blood work was one of the first exercises in biochemistry, the science that uses chemistry to study living systems. He proved that there was iron present in blood. Later it was found that this iron was the vital element that carried oxygen in human red blood cells. Boyle was also the first to use phosphorous to ignite splinters of sulphur-tipped wood – the precursor to commercial matches.

In 1669, Katherine's husband died. Brother and sister continued to share the house in Pall Mall. Boyle was still only forty-two and various people made attempts to find him a wife but to no avail. He was by now a confirmed bachelor, dedicated to his scientific work. He had a stroke in 1670, which left him paralysed initially but he recovered his health and lived for many more years. He died in 1691, just seven days after the death of Katherine, who was thirteen years older.

Strangely, for a man who is ranked among the greatest

scientists of all time, Boyle's exact burial place in London is unknown. It is known that he and his sister were buried in the church of St Martin-in-the-Fields in Trafalgar Square but when the church was demolished in 1721, because of decay found in a survey of its roofs and walls, their bodies were moved and reburied in an unknown plot. There is, however, a commemorative plaque in the crypt of the 'new' neoclassical church that was completed in 1724 and stands on the site today.

The Legacy of Robert Boyle

- Boyle was a key figure in establishing observation, data collection and experimentation as central to the modern scientific method.
- He formulated a 'mechanistic' view of the world, which held that the earth was like a giant machine, composed of lots of moving parts.
- He provided the experimental proof for what is called 'Boyle's Law', which states that pressure and volume are inversely proportional in a gas.
- He was co-founder of the Royal Society, the first society in the world that promoted the 'new science' of observation and experiment.
- He wrote the first English textbook describing electricity.

14

DARWIN'S DEFENDER:

John Tyndall (1820-93)

Birthplace: Leighlinbridge, County Carlow

It is Wednesday, 19 August 1874. John Tyndall, from rural County Carlow, a scientific giant of the Victorian age, is about to get to his feet and make a headline-grabbing speech to his peers at the annual meeting of the British Association for the Advancement of Science (BAAS) in Belfast. This is a confident man – some might say belligerent – a superb communicator, whose public lectures on scientific topics in London are famous. If even the great Tyndall is nervous before this speech he has good reason. He knows its importance and his enemies, primarily from what might be called the 'old guard' of science, will be listening. These are the gentlemen who for centuries have been able to reconcile science and religion. To them, there is no religious-scientific divide or need for rancour or conflict between the two.

The old guard was outraged when Tyndall and his friends at the notorious X Club (more of them later) began a public campaign in support of Charles Darwin's evolutionary theory of man as outlined in his book, *The Origin of Species*, published in 1859. Darwin was not inclined to defend his theory in public, so others had to do it for him. Scientists like the zoologist Thomas Huxley argued that Darwin should at least be given a fair hearing but Tyndall, in this speech, plans to go further. He is well aware that his comments are likely to shock many of those who have come to hear him speak or read about them in newspaper reports all over the world.

The old guard has long been worried about the emergence of a new breed of scientific men like Huxley and Tyndall, who now, at the age of fifty-four, holds the position of President of the esteemed BAAS, a role that provides him with a platform for his views and his agenda, as the President's annual address is always reported with interest by members of the press. The old guard is right to worry about Tyndall, as he represents not just a threat to their outdated ideas and methods but to their waning power base in science.

In his speech, Tyndall plans to shake them further by demanding that religion and superstitions of all kinds, as he sees them, be completely removed from the scientific arena. The reason the sky is blue, why the earth stays warm and is full of life or why there can be wild and seemingly random fluctuations between the earth's warm and cold periods – these can all be explained by rational scientific investigation, observation and experimentation and the answers have nothing at all to do with God or with divine intervention. Pointedly, he will deliver this unequivocal message in his native Ireland, in Belfast, a city famous for its religious fervour. He will further articulate the view that churchmen, particularly the Catholic clergy, have had far too much influence on the education of young minds and that the Catholic Church has stymied the development of science over many centuries. Tyndall's speech will leave his audience gaping and be debated around the world. It will mark the opening salvo of a religion versus science conflict that continues to rage to this very day.

In a moment we will return to Belfast and Tyndall's address but before we do so, it is important to understand the context of his ground-breaking speech. To do that, we must look at the origins of the famous – infamous to some – 'X Club' of 19th-century British scientists, of which Tyndall was a leading member. This private dining club, which was formed in November 1864, had nine founding members, all linked by their 'devotion to science, pure and free, untrammelled by religious dogmas'. As well as sharing the aim of resisting clerical interference in science, as they saw it, the members wished to put science and scientists on a professional footing.

Who were the members of the X Club? Along with Tyndall, they were Thomas Huxley (Tyndall's friend since they met in 1858 and the club's main initiator); Edward Frankland (a chemist who went to school with Tyndall in Queenswood and to Marburg University with him to take a PhD); Thomas Archer Hirst (a mathematician and friend of Tyndall since the 1840s); Joseph Dalton Hooker (a botanist and Charles Darwin's closest friend); George Busk (a naval surgeon, zoologist and palaeontologist); William Spottiswoode (a mathematician and physicist); John Lubbock (the youngest club member by far, a banker who helped establish archaeology as a science); and, lastly, Herbert Spencer (a philosopher, biologist and the man credited with coining the phrase 'survival of the fittest' in a paper he wrote after reading *The Origin of Species*.)

The talented, ambitious members of the X Club challenged the scientific establishment. In some ways, they were outsiders, like Tyndall, who was Irish and whose family had modest financial means. He had received a vocational education in drawing and surveying and paid his own way to do a PhD. Like the other members of the club, Tyndall had to work hard for everything he achieved, something that was in itself a challenge to the established order. Many of the grandees of British science, those who held leadership positions and had done so for centuries, were either churchmen or wealthy amateurs, with leisure and money to pursue their interests. The 'X Club', on the other hand, comprised men from the growing middle classes, who regarded science as their profession and livelihood, not just as a hobby.

The struggle between the forces of 'old' and 'new' for the control of British science and the influential scientific body, the BAAS, had been brewing since the early decades of the 19th century and the covert power struggle emerged into the open following the publication of *The Origin of Species.* The establishment was outraged by Darwin's claims and attacked him at every opportunity. Many of those who came to Darwin's defence, such as Tyndall and Huxley, were later to join the X Club. This group emerged victorious from the struggle, which reached its climax in the early 1860s but simmered for decades afterwards.

By 1868, almost a decade after Darwin's book was published, the X Club had seized control of the BAAS. Between 1868 and 1881, when the club was at the height of its powers, five of its nine members held the BAAS presidency. Therefore, when Tyndall got to his feet in August 1874 to deliver his presidential address, he was doing it from a position of strength. He knew his argument had won the day among his scientific peers, although enemies still lurked within the organisation. The main purpose of his speech was not to win over his fellow scientists but to attack the enemies of the X Club outside the scientific community and to win over public opinion.

In a long and wide-ranging address, Tyndall outlined the entire history of scientific thought as he saw it and the way that its progress was impeded by the power of the Church. He described how Galileo was compelled to swear on the bible that his heliocentric theory – that the earth revolved around the sun – was false, or be burnt alive. The Church's suppression of science was a theme he brought up again and again.

He attacked the theory of spontaneous generation of life – believed by many Christians in the 19th century – which held that life could simply burst into being. Tyndall had found a way to sterilise a surface and kill all life and used this experiment to show that life could not simply grow from nothing on a sterile surface. This was a direct attack on the creationist view of life.

Finally, more than halfway through his address, he reached the main subject of his talk – Darwin and the many reasons he, Tyndall, believed the Darwin's ideas on the origins of man rang true: that the theory fit with much modern scientific thinking and that it was the result of years of painstaking observations by a very thorough and dedicated scientist. He said of Darwin: 'He moves over the subject with the passionless strength of a glacier; and the grinding of the rocks is not always without a counterpart in the logical pulverisation of the objector.'

The conflict between the 'new science' and religion was a hot topic in the mid 1870s and had been ever since Darwin's published his famous book. Many daily newspapers in Britain, Ireland, North America and Europe carried a report of Tyndall's address on their front pages. It caused such a stir that the opponents of Darwin felt obliged to respond, also in the newspapers. Tyndall, as Darwin's most articulate defender, had reached a global audience. This case for 'rational science' had been made and reported around the world.

Without question, Tyndall was one of Ireland's greatest scientists and one of the greatest scientists of his age. His interests were wide-ranging and his influence was huge. He achieved fame in an era (1850-1890) when the United Kingdom of Britain and Ireland was at the peak of its global power and home to many original scientific minds. Yet many Irish people have never heard of Tyndall and this may partly be due to the fact that his background and beliefs were Unionist and anti-Catholic, opposed to the views of the majority of Irish people of his time and today.

An opinion piece Tyndall wrote for *The Times* and which was published on 27 December 1890 gives an insight into his political thinking. He described priests and Catholicism as 'the heart and soul' of the Home Rule movement and said that placing the non-Catholic minority under the dominion of 'the priestly horde' would be 'an unspeakable crime'. He even tried, without success, to get the BAAS to denounce Home Rule as being against the interests of science.

Tyndall was born in 1820 in Leighlinbridge, County Carlow, a small town on the edge of the area known historically as 'the Pale'. For several centuries there had been cultural tensions in the area between newcomers, mainly from England (termed 'settlers' by the English and 'planters' by the locals) and the indigenous people. Tyndall was part of the settler group, as his ancestors had arrived from Gloucestershire in the 17th century, not long after the ruthless Cromwellian conquest of Ireland. It is not hard to imagine why Tyndall was airbrushed out of Irish history, despite his great achievements, particularly after independence was achieved in 1922.

The Tyndalls were Quakers, a dissenting group that broke away from the Church of England in the 1650s, in an attempt to restore what they believed were the authentic practices of the early Christian church. Tyndall's father, also called John, was an intelligent man, of modest means, an RIC constable who worked hard to provide the best possible education to his son. John senior was also, reportedly, fiercely anti-Catholic, something that must have influenced his son. His wife, Sarah McAssey, was descended from wealthy local Catholic landowners but it seems her grandmother, John Tyndall's great-grandmother, had been deprived of her inheritance for marrying a Quaker. Tyndall junior had one sister, Emma.

In 1831, the English government established, in Ireland, the world's first state-supported school system. It was in theory non-denominational – 'to unite in one system children of different creeds', according to the letter from Lord Stanley announcing the initiative. Tyndall, as a Quaker, was part of the first generation of Irish children, along with neighbouring Catholic children, to benefit from this new state school system. He attended the one-room Ballinabranna Mixed National School, which is located half-way between the towns of Leighlinbridge and Carlow. The school is still in existence today, with one hundred and sixty boys and girls on its rolls. Tyndall senior was anxious that his son attend Ballinabranna, despite the fact that he would be mixing with Catholic children, as the master there, John Conwill, also a Catholic and a former hedge-school teacher, was an educator of renown. He made this decision despite fierce pressure from his Protestant neighbours, including the rector of Leighlinbridge, Dean Barnard Boyle, not to send his son to the school. He is said to have remarked that 'even if he [his son] was taught on the steps of the altar', he would send him to Conwill. Tyndall studied English, logic, bookkeeping, drawing, surveying and mathematics, excelling at the two last subjects.

Tyndall's education proved crucial to putting him on the right path: after leaving school, at the age of nineteen, he got a job with the Irish Ordnance Survey and moved to Youghal. There he remained for a few years, until he was chosen to join the more highly regarded English survey and moved over to Preston, Lancashire. It is interesting to note that Tyndall became very unhappy in Preston, as a result of what he regarded as discrimination against Irish assistants in the survey office there. He was so unhappy that he lodged a formal protest on behalf of his fellow-Irishmen about the way in which the assistants were being exploited. He was soon seen as a leader of malcontents and this led to his dismissal in November 1843. Even at the age of twenty-three, Tyndall showed that he was prepared to fight for the rights of others and take on powerful establishment forces, even if it meant losing his job.

Tyndall's next employment was as a railway surveyor on the West Yorkshire Line, during the height of railway fever and the enormous expansion of the railway network. For the next few years he earned an excellent salary but in 1847, to the surprise of his family and friends, he left his lucrative surveying post to take up a teaching job in a progressive Quaker school, Queenswood College in Hampshire. The move showed that money was not Tyndall's god and that his true interests lay in mathematics, as well as the – relatively new, at the time – sciences of physics and engineering. He didn't last long in Queenswood and to take his education a stage further, he left for Germany, in 1848, with a friend, Edward Frankland, the superintendent of the science laboratory in Queenswood. They had both decided to attend Marburg University, a centre of modern scientific thinking, in order to attain a PhD. In Marburg, Tyndall excelled and got his PhD in just two years. The move to Germany was self-financed and the two young men had to suffer some severe hardships in order to eat and have a roof over their heads while they studied.

After he finished his studies in Marburg, Tyndall began to search for an academic job. He applied, without success, for positions in far-flung places like Toronto and Sydney, as well as Cork and Galway. A major turning point came in February 1853 when he was invited to give a Friday lecture at the famous Royal Institution of Great Britain (RI) in London. The RI, which is devoted to scientific education and research, was founded in 1799 by a group of leading British scientists. Michael Faraday, the inventor and pioneer in electricity, was its head.

Tyndall's lecture impressed everyone who heard it, including Faraday, who invited him to give further lectures. Quickly, he became famous for his ability to explain difficult concepts to a non-scientific audience. The crowds came to hear him and his fame rapidly grew. A few months later, in May, he was unanimously appointed Professor of Natural Philosophy

(Physics). The agreement was that he was to provide nineteen public lectures per year for a salary of £200. This was a substantial sum, considering that he was committed to giving only one to two lectures per month, at a time when the average salary for a clerk was about £150 per annum. Tyndall was thirty-three years old, his career was finally on track and he was on his way to becoming an influential figure in the world of science and beyond.

Tyndall spent the rest of his working life in RI, succeeding Michael Faraday as director when Faraday died in 1867. He retired in 1887, aged sixty-seven, after thirty-four years working there. He left his mark on the institution, particularly in the way he continued the efforts begun under Faraday to popularise science and make it intelligible to the public. Tyndall was, by all accounts, a gifted lecturer and communicator, who used his oratorical skills and experimental brilliance to hold an audience, often leaving them spellbound at the end. He could reasonably be called the first professional populariser of science and was certainly one of the first science writers ever to reach out to a broad audience. As well as lecturing to the public in the RI, Tyndall went on tours overseas, to the US, for example, and wrote many books, explaining scientific concepts. His lectures always attracted a crowd and his books were so widely read that he became a wealthy man.

After his appointment in the RI, Tyndall was hungry to get to work. He wanted to explore the wonders of nature. By the late 1850s he had become focused on understanding the radiant heat of the sun and its relationship with the various gases present in the earth's atmosphere. These investigations would lead to what was arguably his greatest single contribution to science: the discovery of what we now call greenhouse gases – gases that retain the heat of the sun, keeping the earth warm and making life possible here through the 'greenhouse effect'.

Tyndall discovered that two gases present in tiny or trace concentrations in the atmosphere were absolutely critical for life. These key gases are carbon dioxide (CO_2) and water vapour, which between them constitute less than two per cent of the gases in the atmosphere. Tyndall demonstrated that CO_2 and water vapour could effectively absorb and retain the heat of the sun, unlike far more common atmospheric gases such as nitrogen and oxygen, which comprise more than 98 per cent of the gases in the atmosphere. The importance of his finding was that without CO_2 and water vapour, which are now called the greenhouse gases, the heat of the sun would travel through the earth's atmosphere to the surface, where much of it would bounce back up and travel unimpeded back out to space. None of the sun's precious heat would be trapped and earth would be nothing more than an icy, lifeless, lump of rock. These days, greenhouse gases have a negative association with relentless global warming but without them, no life could exist here on earth.

Tyndall is credited with being the first to explain and demonstrate why the sky is blue. In one of his most famous lectures, he set up an experiment that stunned his audience by artificially recreating blue skies inside the lecture theatre of the RI. What he did first was to show how light, which appears white, normally travels in a straight line unless it is blocked, reflected or absorbed by something. That 'something' can be fine particles in the atmosphere, such as dust, pollen or even salt from the oceans, or it can be atmospheric gases.

The creation of the artificial sky inside the lecture theatre required that some of the white light travelling in a straight line be absorbed and radiated. White light, of course, is composed of all the components of the rainbow when split into its various parts. Light travels in waves. Different wavelengths, which are defined by the interval between wave repetitions, produce different colours. Red, orange and yellow are produced by longer wavelengths and these tend to pass through the atmosphere unimpeded. However, the shorter wavelengths, which produce the colour blue, tend to be absorbed by atmospheric gases. The absorbed blue light is then radiated or scattered all over the sky, giving the sky its blue hue. Tyndall's work inspired Lord Rayleigh to develop a measurable, mathematical explanation of why the sky is blue, which famously became known as 'Rayleigh scattering'.

Tyndall's work on light had two further significant developments. One had to do with measuring whether air could be made free of bacteria and germs and the other led to the development of a test for whether it was possible for life to arise spontaneously, which creationists then believed (some still believe it today).

In numerous experiments in the early 1870s, Tyndall showed that 'optically pure air' was free of bacteria and germs. At around the same time, Louis Pasteur was postulating that living germs floating in the atmosphere were a cause of human and animal disease. Tyndall was able to demonstrate that foods exposed to optically pure air after boiling remained unspoiled, whereas the same foods, when exposed to ordinary dust-laden air, swarmed with bacteria.

Tyndall wanted to show all rational people that creationism was irrational and unscientific. At the time the main advocate of the creationist argument was a London pathologist called Charlton Bastian, who was a fellow of the prestigious Royal Society, a group of leading scientists and researchers. Tyndall was determined to prove him wrong. In 1877, he came up with a process of repeated heating, which he called discontinuous heating, which succeeded in sterilising liquids containing the most resistant germs. This method became known as 'tyndallisation' in France and 'pasteurisation' in Britain. Tyndall then sterilised the inside of a box, left it for some time untouched, came back, unsealed it and found no living thing residing inside. This was a simple proof that life could not arise out of nothing.

Tyndall was also the first to design experiments that 'guided' light, using the forerunner to fibre optic cabling that is so crucial for high-speed internet communications in many modern advanced industries. His achievement in being the first to produce optically pure air resonates in his homeland to this day, as many of Ireland's advanced manufacturing units, such as Intel in Leixlip, are dependent on cleanrooms, where air is ultra-clean, preventing even the tiniest of particles from contaminating the manufacturing process. It can be strongly argued that cleanrooms began with Tyndall.

Tyndall was what we would call today a 'workaholic' and his output, in terms of books and letters, experimental work and lectures was phenomenal. It is remarkable in this context that he succeeded in having a private life. In his twenties and thirties he had a number of relationships with women but it was only in 1869, at the relatively advanced age of forty-nine, that he made his first proposal of marriage. It was rejected. After that rebuttal, it appeared certain to Tyndall's friends and family that he would remain a bachelor for life, yet he surprised them all when he announced five years later that he was to marry Louisa Charlotte Hamilton, some twenty-five years his junior. Louisa was from aristocratic stock, daughter of Lord Claud Hamilton and Lady Elizabeth Proby. She was also a very committed Christian. Despite their differences, it appears that theirs was a very happy marriage. The couple did not have any children.

Tyndall become very interested in mountaineering after travelling to the Alps in his mid-thirties to study the mechanical behaviour of glaciers. He was an excellent climber, first to reach various Alpine summits. Tyndall's mind was always working and his observations of glaciers while in the mountains led him to write a book, *The Glaciers of the Alps* (1860). In 1877, the Tyndalls built a summer cottage in the Alps, in Brig, southern Switzerland, not far from the Italian border, and they spent time there every year from then on.

Tyndall suffered from insomnia and dyspepsia (indigestion and stomach upset) almost all his life but the problems got worse as he got older and he also suffered from attacks of phlebitis (vein inflammation). Illness let directly to his retirement in 1887, aged sixty-six. Unlike other Victorian men of science, Tyndall had not inherited a fortune, so it was fortunate for him that he amassed a great deal of wealth from his lecturing and writing throughout a hugely successful career.

The manner of his death was tragic and bizarre. As his health deteriorated in his seventies, Tyndall regularly took magnesia at night, something considered beneficial for overall good health. However, on one particular occasion, his wife gave him an overdose of chloral hydrate instead of magnesia. Chloral hydrate is a sedative that is also used as a chemical reagent, a substance that is added to a system, such as a fluid, to cause a chemical reaction, or determine whether a chemical reaction has occurred. The effect of chloral hydrate poisoning is to attack the central nervous system, leading to vomiting, stupor, coma and death. This is what caused Tyndall's death on 4 December 1893, at the age of seventy-three.

Tyndall realised what had happened, expressed sympathy for his wife's plight and said goodbye. An inquest absolved the distraught Louisa of all blame. She had devoted herself to Tyndall, as personal assistant and nurse. After he died she spent years gathering information to write his biography, which was finally published in 1945, five years after her death at the age of ninety-five.

The Legacy of John Tyndall

- He is considered the father of meteorology and climate science as he discovered the presence of greenhouse gases in the earth's atmosphere.
- He defended Charles Darwin against many detractors, after the publication of *The Origin of Species* in 1859.
- He developed 'optically pure air' – the forerunner of clean-room technology, which is vital for today's advanced manufacturing industry.
- He was the first to devise a way to 'guide' the direction in which light travels. He developed a light box – the precursor of modern fibre-optic cabling.
- He was one of the greatest science writers, lecturers and popularisers of the Victorian era, making science intelligible to a wider audience.
- He helped establish science as a profession rather than a hobby for wealthy amateurs and churchmen, as it had been up to the 19th century.
- He was a member of the X Club of leading Victorian scientists who demanded that science be free from the influence of religious dogma and bias of all kinds.

15

WHAT IS LIFE?:

Erwin Schrödinger (1887-1961)

Birthplace: Vienna; became a naturalised Irish citizen in 1948

It is February 1943. The five-month battle for Stalingrad, one of the bloodiest in human history, which has claimed an estimated two million lives, is coming to an end. The suffering of people all over Europe and around the world has been incalculable and the Second World War still has more than two years to run. Meanwhile, in a hushed lecture theatre in TCD, in neutral Ireland, a renowned, Austrian-born man of science, who has found refuge from the Nazis in Ireland, gets to his feet. The man is fifty-six-year-old Erwin Schrödinger, one of the most famous physicists in the world, whom some people consider to be on a par with Albert Einstein. There is an audience of about four hundred, including Taoiseach Éamon de Valera and members of his cabinet.

The subject matter of today's lecture and the two that are planned to follow is way outside Schrödinger's usual area of expertise and he is risking criticism from fellow scientists by addressing it. Schrödinger, who will become a naturalised Irish citizen five years from now, plans to apply the laws of physics to the greatest of all biological questions: what is life? Or, more precisely, what is the stuff that determines the form life takes and how is this information passed between the generations? The lectures, which will be published in 1944 in a book called *What Is*

Life?, will inspire young physics-minded scientists to switch their focus to biology. Ten years later, three of them will succeed in describing the structure of deoxyribonucleic acid – (DNA) – the code for life. But more of this later.

Dublin, 1939. A drab, overcast, economically depressed, slowly decaying city, isolated by war, living on rations and in a perennial state of high anxiety. This is a town famous for the genius of its writers but in terms of science, there is nothing – it's a desert. Erwin Schrödinger has not decided to come here because of the presence of other great scientific minds, or for the resources that will be put at his disposal. He has come because he has been offered a job by Éamon de Valera, which provides him with something that the cosmopolitan cities at the centre of Europe can't hope to match – sanctuary from the Nazis.

For Schrödinger, this safety comes at a price. As a scientist of the highest rank, in Dublin, Schrödinger will be intellectually isolated from other members of the same species. Some top scientists do manage to visit but it's wartime and the seas are dangerous so visitors are scarce. Schrödinger will suffer because he won't have another gifted thinker down the road with whom to discuss things on a daily basis; nor will he have ready access to scientific journals and the latest research they report. European journals arrive in Ireland by mail boat service, which can be disrupted by U-boats on patrol in the Channel. It is not surprising, then, that Schrödinger's mind wanders from his specialist area – theoretical physics and the problems of defining the tiny quantum world – into other aspects of science and of life. Hence his idea for a series of lectures in TCD devoted to the question: 'What Is Life?'

Schrödinger's scientific stature has nothing to do with biology and genes and everything to do with physics and the emergence in the 20th century of the field of quantum physics, which outlined the laws of behaviour for tiny particles. In the early part of the 20th century, experimental physicists had begun to see a range of strange results from their experimental observations of the quantum world. The evidence appeared to show that atoms, or even smaller sub-atomic particles, did not obey the classic laws of physics, as outlined by Isaac Newton in the 17th and early 18th century. Newton had described the motion of celestial bodies and the laws of gravity. These classic laws worked very well for planets moving through space and for large objects moving here on earth but they were of no use for describing the movement of the tiniest particles.

The experimentalists were baffled by what they were seeing. For example, electrons, the sub-atomic particles responsible for electricity that spin around the nucleus or centre of an atom, were found to be existing in two states at once. Trying to understand this was akin to trying to understand how a light switch can be turned on and off at the same time. For a time there was no coherent theory to explain this weirdness. In 1926, Schrödinger was the first to put forward a proposal, in the form of the Schrödinger Equation, that could help explain the oddness observed in the quantum world. The equation stood up to lab testing and helped lay the foundations for a new field of physics called quantum mechanics, which sought to describe the atomic world accurately. Schrödinger, along with Paul Dirac, won the 1933 Nobel Prize for Physics for his equation. By the time he arrived in Dublin in October 1939 Schrödinger was one of the world's most famous scientists. To get him here was a huge coup for science in Ireland.

The circumstances of Schrödinger's arrival in Dublin late in 1939, to join a tiny band of some fifty-two expatriate Austrians, had been, like many aspects of his life and work, extraordinary. This is a story of intrigue, danger and drama that begins with Éamon de Valera's vision for an Irish Institute of Advanced Studies, which would incorporate advanced physics, mathematics (de Valera was himself a mathematician) and Celtic Studies (another of his passions). Once de Valera had formulated his vision for what later became the Dublin Institute for Advanced Studies (DIAS), he and his advisers, academics like T.W.T. Dillon, Professor of Therapeutics and Pharmacology in UCD, were keen to get a big name to head up the new institute. Professor Dillon had spent time studying modern medicine in Vienna before returning to take up his post in UCD. It is likely he was aware of the work of the Vienna-born Schrödinger and the fact that he had run into trouble with the Nazis while working in Berlin.

Schrödinger's star had risen quickly following the publication of his equation. In 1927 he was appointed Professor of Theoretical Physics in the Friedrich Wilhelm University in Berlin, in succession to the great German scientist, Max Planck. This was a prestigious, career-enhancing appointment for Schrödinger. All went well for him until the Nazis came to power in 1933, when he opposed the Nazi practice of publicly burning books written by Jewish authors as well as the appointment of party sympathisers to important academic posts. The Nazis, noting his opposition, had him in their sights. He felt threatened enough to slip out of Germany on the quiet and head for England, to the University of Oxford.

Schrödinger stayed in Oxford for a while but things didn't work out for him there. It is not entirely clear why this was the case but it appears that part of the reason was unease about Schrödinger's unusual domestic arrangements, which involved his wife, Anny (Annemarie Bertel), Hilde March, the wife of fellow Austrian physicist, Arthur March, who was his mistress, and Hilde's daughter. When Oxford did not work out, he was offered a post in an American blue-chip university, Princeton, but his living with two adult women also proved a stumbling block for Schrödinger in the US. In 1936, he took a gamble – given the threat to Austria posed by Nazi Germany – and returned to his own country to take up a post as Professor of Physics at Graz University. The gamble failed when Germany annexed Austria in 1938. Once again Schrödinger found himself at the mercy of the Nazis. They had not forgotten his opposition to them in Berlin and promptly fired him from his job at Graz. It was clear that this was just a first step and that Schrödinger and his wife were in real physical danger. Again he fled. This time he managed to get to the neutral Vatican with the help of Italian physicist, Enrico Fermi, who had Schrödinger admitted to the Pontifical Academy of Sciences.

De Valera now entered the story. He heard about the trouble Schrödinger was in and began a secretive process of making contact with him in order to lure him to Ireland. Otto Glaser, a fellow Austrian expatriate in Ireland, described what happened next, as told to him by Anny Schrödinger. De Valera got in touch first with the English mathematician, Sir Edmund Whittaker, who was a convert to Catholicism and also a member of the Pontifical Academy of Sciences. It is likely that de Valera met Whittaker during his time as Royal Astronomer of Ireland between 1906 and 1911.

Whittaker in turn got in touch with Max Born, a German-British physicist who was also prominent in the birth of quantum mechanics. Born wrote to a Professor Richard Bar, a physicist and mathematician based in Zurich who was a friend of both Born and Schrödinger. Bar asked an unnamed Dutch friend, who happened to be on his way to Vienna – Schrödinger's hometown – to drop a message to Anny Schrödinger's mother there. The Dutchman agreed, visited Anny's mother, handed her a slip of paper, said it was from Bar and left. Anny's mother passed the note on to Schrödinger in a letter. This could have been a fatal mistake, as the Gestapo were by now controlling Austria and reading letters coming in and out of the country. But the note reached Schrödinger, who read it and destroyed it. It stated that

Mr de Valera planned to found an Institute for Advanced Studies in Dublin and wanted to know whether Schrödinger, might, in principle, be interested in being its first director.

Schrödinger was interested and arrangements were made for him meet de Valera in Geneva, where the Taoiseach was working in his capacity as President of the League of Nations. On his way Schrödinger had to endure a nerve-wracking moment at the Swiss-Italian border when suspicious Swiss border guards thought it was odd that a first-class passenger was travelling so light. They held him up, before finally letting him pass. In Geneva, de Valera and Schrödinger met and agreed in principle that Schrödinger would head up the new DIAS once it had been legally established in Ireland. This would take a bit of time, so in the meantime Schrödinger took a guest professorship in Belgium. De Valera finally sent for Schrödinger in October 1939 and arranged for his safe passage through England to Dublin.

To those who knew Schrödinger, like Otto Glaser, a successful businessman now living in Howth, County Dublin, it is not surprising that he decided to address the question, what is life? Otto remembers a vibrant man, full of energy, full of life and ideas and interested in many things. It made sense that Schrödinger, a gifted man with time on his hands, might focus on that question, one of the biggest facing the whole of science, while he was working in Dublin.

The idea of tackling this issue seems to have been Schrödinger's own and it was a brave move. The lectures were sure to attract controversy and most scientists, worried about the impact on their careers, would avoid speaking on topics outside their immediate area of expertise. If Schrödinger had talked to other scientists in Ireland about his plans, it seems likely that physicists would not have been interested as they would regard the question as outside their remit, while geneticists didn't yet exist in Ireland. The first genetics department in Ireland was not established until 1958 (in TCD). The fact that Schrödinger developed his ideas of life in this intellectual vacuum makes them all the more impressive.

Schrödinger knew that no matter what people thought of the ideas he put forward on the 'What Is Life?' question, his reputation in theoretical physics was assured forever because of his formulation of the Schrödinger Equation. Nevertheless, what he was doing was intellectually courageous. He was moving outside his scientific comfort zone, something that was simply not done in the 1940s. Even today, few scientists would attempt it. It is too easy for specialists in a particular area to dismiss what is said about their own area by outsiders, even other scientists, on the basis that the outsider knows nothing about that area. Schrödinger was aware of the likelihood of criticism from other quarters but went ahead regardless. He believed that he could make an important contribution to the 'What Is Life?' debate by applying physics to the problem.

Schrödinger's three lectures on the subject followed a basic principle of science, which is a key component of all types of biological research today: if the structure of something can be determined, finding its function becomes a lot easier. By entering the 'What Is Life?' debate in such a dramatic fashion, Schrödinger was issuing a challenge to anyone who would listen, saying that genetic material, whatever it may be, must have particular properties. He listed these properties from the point of view of general physical laws and challenged scientists to find the material.

In February 1943, as Schrödinger got to his feet to address his audience in TCD, genetics was in its infancy as a scientific discipline. There were only a few places it was taken seriously, such as California Institute of Technology, Columbia University, New York, and Cambridge University. In these universities teams of researchers were in place who were interested in finding out more about the code for life and how it worked. The series of lectures that Schrödinger was about to deliver would prove to be a huge boost to the field.

The importance of the book *What Is Life?* that was published a year later lay in how it inspired a generation of emerging physicists to focus not on questions of physics but on finding the code for life. Amongst this group were Francis Crick, James Watson and Ireland's Maurice Wilkins (*see* Chapter 16). In 1953, they made the most important biological discovery since Darwin outlined his theory of natural selection in 1859, when they identified the precise structure of DNA, the magical material that held the code for life. They all said they had read *What Is Life?* and that it had proved to an inspiration.

After Watson and Crick reported their discovery in *Nature* on 25 April 1953, the two men were the subject of global media attention. They became celebrity scientists and their lives changed forever. But Crick found time to write a thank you-letter to Schrödinger on behalf of himself and Watson, addressed to him at his home, 26 Kincora Road, Clontarf:

'Dear Professor Schrödinger, Watson and I were once discussing how we came to enter the field of molecular biology and we discovered that we had both been influenced by your little book, *What Is Life?*'

Maurice Wilkins, the so-called 'Third Man of the Double Helix' (the title of his 2003 autobiography), also said that he was strongly influenced by *What Is Life?* After the Second World War, Wilkins decided to become a biophysicist, using physics to solve biological problems, such as finding DNA's structure. In his autobiography Wilkins wrote: 'Schrödinger used the language of physicists and that stimulated me, as a physicist, to persevere with his book and its introduction to genetics and to decide that this was the general area that I wanted to explore as a biophysicist.'

One of the key concepts that Schrödinger introduced in *What*

Is Life? was the idea that the genetic material, whatever it is, must be what he called an 'aperiodic crystal'. This term and the idea behind it had a lasting impact on Watson, Crick and Wilkins.

Crick's 1953 letter to Schrödinger concluded: 'We thought you might be interested in the enclosed reprints [of their ground-breaking DNA paper] – you will see that it looks as though your term "aperiodic crystal" is going to be a very apt one.'

Wilkins noted in his autobiography:

'He [Schrödinger] wrote about a gene being an aperiodic crystal and that connected directly with my PhD research where electrons moved freely in perfect crystals but could be slowed down and trapped when the crystal had irregularity. It seemed to me that 'aperiodic' referred to the local irregularities in which the genetic message was written against a periodic background.'

When Schrödinger used the term 'aperiodic' to describe the genetic material, what he was essentially saying was that any genetic code must contain variation, as well as having elements that do not change. By 1943, DNA had been identified as one of the molecules that was a strong candidate for housing the genetic code. Schrödinger knew this and he also knew that DNA was a salt. A salt can be dissolved in water and it forms crystals when the water is heated and evaporated. Thus his use of the term 'crystal' when talking of the genetic material.

Crystals have an orderly structure, which is a good thing for any potential genetic material, but it is only a starting point. Order is required, as without it information cannot pass in a useful manner. However, as well as having a high level of order, Schrödinger said the genetic molecule must contain disorder. The reason is that if a molecule is completely orderly, with no variation, it contains no information. For example, snowflake crystals would be useless as a genetic material: there are countless possible snowflake crystal types but within a batch, one snowflake is the same as the next. Once the conditions are right for a snowflake to form, a snowflake crystal of a certain type is produced and all the other crystals in the same batch will be identical. There is no variation. A snowflake is, therefore, a periodic crystal, akin to a line of repeating letter As.

What Schrödinger did for the emerging group of biologically-inclined physicists was to provide them with a checklist of the necessary properties of the genetic material. He said that the material had to be easily able to reproduce itself, had to contain variation, had to have order and also had to be capable of mutation or dramatic change. He provided a road map for people who were desperate to find the stuff of life. When DNA was finally described, its properties were consistent with those outlined by Schrödinger in TCD in 1943: this was one of the reasons Crick wrote him a thank-you note in 1953.

Schrödinger arrived in Dublin on 7 October 1939, with Anny, Hilde and Hilde's daughter Ruth – his wife, his mistress and his mistress's daughter. It was two weeks since Warsaw had been flattened by the Luftwaffe and two days since the Polish army had surrendered to the Nazis. It is interesting to note that in the über-Catholic Ireland of 1939, Schrödinger's unorthodox domestic arrangements caused no problems. These had proved too much for people in the liberal, intellectual environs of Oxford and Princeton to accept but in supposedly 'backward' Ireland, people either turned a blind eye or didn't take a blind bit of notice, it seems. While in Dublin Schrödinger fathered children with two different women but again there was no adverse reaction to behaviour that would normally have been viewed as totally 'beyond the Pale'. Éamon de Valera must have believed that the presence in Ireland of Schrödinger, although a free-thinking atheist, was good for the country, despite the complexities of his private life. Either that or the staunchly Catholic Taoiseach simply didn't know what was going on.

Schrödinger kept a log of his sexual adventures and it's

likely that he had children by at least three mistresses. Ruth, the daughter of Hilde March, was his daughter, it seems. It also appears that Hilde's husband, Schrödinger's colleague, Arthur March, had a relationship with Anny Schrödinger. Schrödinger moved in Bohemian circles during his seventeen years in Dublin. He was sociable, engaging, charming, wrote poetry and was interested in the many aspects of the arts, as well as the sciences. He was a regular visitor to the Abbey Theatre, fond of Irish music and learned the Irish language. He was popular with students and invited them to the lively parties he held in his home in Kincora Road, Clontarf. He became a well-known figure around Dublin, cycling to work every day from his home to his office in Merrion Square in the city centre. He liked the outdoors, went hiking in the countryside and was very fit. He said later that he had been very happy living in Dublin.

Schrödinger had to wait for more than a year after his arrival in Dublin, until 21 November 1940, for legislation to be passed allowing the setting-up of DIAS. Only then could he be officially appointed its first director. In the interim, de Valera arranged for him to conduct lectures on wave mechanics at UCD. This was the science describing the dual nature of light – that it was a wave and a collection of particles at the same time – that Schrödinger had invented virtually from scratch. The fortunate UCD undergraduates heard the story straight from the great man himself.

In July 1940, as Schrödinger continued to wait for DIAS to open its doors, both UCD and TCD conferred him with honorary doctorates of science within a week of each other. At the UCD conferring, the president, Arthur Conway, neatly summed up Schrödinger's status in science: 'His name will forever stand out as one of the great names of mathematical physics, taking rank with Newton, Laplace, Hamilton.'

The DIAS had been set up purely to conduct advanced

research and did not have a teaching remit. However, Schrödinger wanted to show his gratitude to de Valera and to Ireland so he devised a number of special lectures for audiences ranging from undergraduates to professors. Otto Glaser, a science postgraduate, befriended Schrödinger at the time. Glaser said that he was a good teacher and found him 'crisp, very clear, anxious to simplify, sometimes humorous and above all desirous to share his undoubted enthusiasm for a difficult subject matter.'

It was good for Schrödinger that de Valera took a keen interest in him and attended many of his lectures. This meant that anything Schrödinger wanted could be arranged quickly, even down to the appointment of a extra office cleaner. His appointment lived up to de Valera's expectations, as his name attracted many top scientists to speak in Ireland and he gathered a distinguished staff of academics around him in DIAS. This established the institute on a firm footing from the beginning and helped to establish its international reputation for research, which endures to this day.

Although Schrödinger loved Ireland, he would probably have returned to Austria earlier had it not been for the political situation. After the war, Austria was divided into four zones, like Germany, with Vienna, like Berlin, also divided in four. Germany was divided into East and West Germany in 1949, as the Cold War escalated, but Austria did not suffer the same fate, continuing to be run jointly by the four allied powers until 1955. After making promises of perpetual neutrality Austria regained its independence on 12 May 1955. The Schrödingers could finally return home and claim their pension rights.

In 1956, Schrödinger returned to Vienna, where he became Chair of Physics at the University of Vienna. He worked there for a few years but died of tuberculosis in 1961. He had suffered from the disease since the 1920s. He had discovered his Schrödinger Equation while recuperating from a bout of illness in a sanatorium at Arosa, eastern Switzerland. His wife, Anny, died in 1965. The Schrödingers had had a famously open marriage, with both having lovers, but they were apparently devoted to each other and were married for forty-one years. The two were buried together in Alpbach, a beautiful small town in the Tyrol.

The Legacy of Erwin Schrödinger

- He devised the Schrödinger Equation, which accurately described the strange nature of the sub-atomic quantum world for the first time.
- His 'What Is Life?' lectures in Dublin in 1943 inspired Francis Crick, James Watson and Maurice Wilkins to find and describe DNA, the genetic code for life.
- He was one of the founding fathers of the science of quantum mechanics.
- He was the first director of the Dublin Institute for Advanced Studies, which was established by statute in 1940.

16

DNA'S THIRD MAN:

Maurice Wilkins (1916-2004)

Birthplace: New Zealand, to Irish parents

It is 12 May 1950 and Swiss biochemist Rudolf Signer has just finished speaking to the Faraday Society at the University of Cambridge. Signer has been working as part of an international team interested in studying the shape and size of deoxyribonucleic acid, also known as DNA. This is a molecule that many scientists are beginning to suspect plays a central role in the passing on of human traits, such as height, eye colour and even personality, from one generation to the next. In keeping with the best traditions of cooperative science, Signer offers pre-prepared samples of his DNA to anyone present who is interested in studying the material closely. A queue forms and in that queue is Maurice Wilkins, a mild-mannered former physicist who worked on the Manhattan Project to develop an atomic bomb during the Second World War but has now turned his attention to finding the elusive biochemical substance that contains the code for life.

Wilkins was born in New Zealand in 1916 and educated in England but, genetically speaking, he was 100 per cent Irish, as his parents, Edgar Wilkins and Eveline Whittaker, were Dubliners. In just a few months Wilkins would produce the first photographic evidence, using x-rays, that the Signer DNA was orderly, with a very clear structure. This finding would prove crucial as serious genetic researchers believed that any material potentially holding the code for life must be highly ordered. Wilkins presented his x-ray photograph to a scientific conference in Naples later in that summer of 1950. It fascinated many present, including Jim Watson, a gifted young American geneticist, and inspired him to seek more detailed proof that DNA and not proteins, as some thought, held the genetic code. In 1952, Watson and his colleague Francis Crick stunned the world by stating that DNA was the code. They produced a model showing what they believed DNA looked like, the famous double-helix structure. For the next decade, Wilkins played a central role, again using x-rays, to confirm in the lab that DNA did indeed look like Watson and Crick said it looked. In 1962, Wilkins, Watson and Crick jointly won the Nobel Prize for Medicine or Physiology for their DNA work. Watson and Crick were colourful characters, articulate and outspoken, which helped seal their lasting fame. The more self-effacing Wilkins, who died in 2004, became the somewhat forgotten 'third man' of DNA.

As far as Wilkins could determine, the Wilkins family had been in Dublin for at least four generations before his father and mother, Edgar and Eveline, emigrated to New Zealand. His mother's family, the Huttons, had been in Dublin for at least a century. One of Wilkins's mother's ancestors, Sarah Hutton, was the mother of the great Irish mathematician, William Rowan Hamilton (*see* Chapter 8), who was born in 1805 in Sarah's home in 38 Dominick Street, in what is now the economically depressed north city centre area of Dublin. It was said that Hamilton had inherited his genius from the Huttons, who also produced Wilkins. The Huttons were a wealthy and influential family in Dublin.

The background of the Wilkins family in the city was humbler. Wilkins managed to trace his family tree back to his great-great-grandfather, who owned a shop on Patrick Street, which runs from Christ Church Cathedral south to St Patrick's. It seems the shop at one point sold candles and at another time it was a butcher's shop. But his great-great -grandfather married a woman with some land and he family moved out to Whitehall, south of Terenure, then a rural area at the end of a tramline. His great-grandfather was a surgeon in the British Army, serving in the 19th-century Afghan Wars, and three of his children graduated from university, including Wilkins's grandfather, William Wilkins, who excelled academically in TCD. He became a teacher and was appointed headmaster of High School (then located in Harcourt Street, now in Rathgar, south of the city centre) in 1879, at the age of just twenty-seven. He married Mary Hutton and the couple lived in 8 Rostrevor Terrace, Rathgar.

In the end the marriage broke down and the Wilkins and Hutton sides of the family didn't speak after this. The breakdown must have been common knowledge around Dublin, as it was even reported in *Ulysses*, where the author, James Joyce, who had been born in Rathgar, commented (apparently favouring Mary) that Mary 'had her work cut out for her' with her stern husband William.

William and Mary had three children: Edgar, Maurice and Una. Edgar, Maurice's father, and Maurice were both pupils at High School when their stern father was headmaster there and the experience seems to have affected them in different ways. Maurice reportedly had a happy time at school and did very well academically. Edgar, it seems, was unhappy and had poor handwriting and his father was ashamed of him. He graduated in medicine in TCD but loved the outdoor life and on one of his adventures in Skerries, north of Dublin, he met his future wife Eveline Whittaker, the daughter of an officer in the Royal Irish Constabulary. Eveline's father belonged to the Plymouth Brethren, an ultra-conservative, evangelical Christian movement that began life in Dublin in the 1820s but Eveline was not so severe, it seems. In his autobiography, *Maurice Wilkins, the Third*

Man of the DNA Helix (2004), the author describes his mother as 'a police chief's daughter from Dublin, an affectionate, beautiful woman with long blonde hair and a great deal of common sense.' Edgar and Eveline married in 1913, sharing a strong interest in medicine, nutrition, vegetarianism and self-learning. Later that year, they emigrated to New Zealand. Edgar was very interested in preventative medicine and it is possible that he saw New Zealand as a new country that might be open to his new ideas.

As things turned out, the couple wouldn't set foot in Ireland for another decade. They headed for Wellington, where Wilkins's beloved sister, Eithne, his only sibling, was born in 1914. The family then moved to the 'wild west' area of Pongaroa, a remote farming region north of Wellington, an area without proper roads, where the locals needed a doctor. Edgar built a wooden house for his family and Wilkins was born in this house, on 15 December 1916.

The family had an active, healthy lifestyle, hiking in the mountains, eating porridge and dancing to John McCormack songs on the gramophone at night. Edgar built up a reputation in the field of preventative medicine and was appointed New Zealand Director of School Hygiene, a job that gave him the chance of starting people down the road of preventative medicine at an early age. It also meant moving back to Wellington. There the family took up residence in Kelburn Parade, a respectable area overlooking Wellington Bay, where Edgar and Eveline named their house 'Skerries' in memory of where they first met. Wilkins had a very happy childhood there. The siblings did not attend school as their parents believed that children should not learn from books but through self-learning.

Edgar became increasingly frustrated that his colleagues were not taking his ideas on preventative medicine seriously. Eveline and Maurice were getting older and needed formal schooling. For these reasons the family decided to leave New Zealand in 1923 and set out on the long trip back to Ireland. But the Dublin that Edgar and Eithne returned to was alien to them. Ireland had won independence from Britain in 1922 but then came the Civil War, which ended only the month before they left New Zealand. The Dublin relatives were warm and friendly, Wilkins later recalled, particular his Granny Whittaker, with whom the family stayed. But the city seemed empty and depressed and jobs were scarce so it was no surprise then that Edgar began to think of emigrating again, this time to Britain. The family did, however, retain a lifelong emotional connection with Ireland, exemplified by the fact that Edgar, who died in 1945, chose to be buried in the same plot as his father, William, in Mount Jerome Cemetery in Harold's Cross. (Despite all his focus on exercise and healthy food Edgar died of heart failure at the age of fifty-seven.)

Edgar decided that his best career prospects lay in England with the British School Medical Service. He spent a year in London, in King's College (where Wilkins later did the work that led to his Nobel Prize), studying for a Diploma in Public Health. He then got a job in a Birmingham school. The family moved to Wylde Green, north of the city, and started a new life there. Eithne was eight and a half and Maurice was six and a half and neither had yet attended school. Maurice gained entry to King Edward's School in Birmingham and later obtained a scholarship to study physics in the University of Cambridge.

As a child Wilkins had built telescopes and little flying aeroplanes in his own workshop. He avidly read about the exploits of people like Major Segrave and Captain Malcolm Campbell, who were breaking the world land speed records for Britain at this time. He had a clear aptitude for science, particularly physics. In Cambridge, John Cockcroft was his tutor (Cockcroft and Ernest Walton received the 1957 Nobel Prize for Physics; *see* Chapter 9) and he was hugely impressed by the work of leading Cambridge scientists like J.D. Bernal, an Irish-born Jew, who was using x-rays to understand biological structures better.

Wilkins became active in the Cambridge Scientists' Anti-War Group while attending university in the 1930s. Many academics of conscience were opposing war, while also opposing the Nazis. Wilkins's involvement in politics caused him to neglect his academic work and he blamed this, as well as periods of depression, from which he suffered throughout his life, for obtaining only a Second Class Honours, Grade Two, in his final degree. This result meant that he would not be allowed to do postgraduate research in the University of Cambridge.

Wilkins wanted to continue in science, despite this setback, and began looking around for options. M.L. Oliphant, one of the big names he had come across in Cambridge, had left to head up the Physics Department in the University of Birmingham, where he was building Britain's biggest 'atom smasher'. Wilkins had decided he wanted to study luminescence in solids, to find answers to the questions of how and why solids emit light when they are heated. He approached Oliphant, who put him in contact with a man called J.T. Randall, who was looking for a research assistant in this precise area.

This was a real stroke of luck for Wilkins. He found himself working in a dynamic lab, very much connected to the real world and its problems, and had the satisfaction of contributing to the British war effort at an early stage. His PhD research helped to explain why radar screens were not clear and was part of a general effort to develop radar, which proved crucial to the success of the RAF in the Battle of Britain.

As the war dragged on into 1941, it became clear that neither side would have a speedy victory. At the end of that year the US entered the war and some scientists began thinking about the possibility of making an atomic bomb using uranium. Most physicists believed that a lot of uranium would be required to make an A-bomb but Otto Peierls and Rudolf Frisch, who both worked in Oliphant's Birmingham lab, calculated that, in fact, not as much uranium was needed as people had thought. This had enormous implications: if an A-bomb was easier to build than people had thought, the Nazis might realise this too. This led to the birth of the Manhattan Project, which resulted in Oliphant's entire lab, Wilkins included, being transported out to the US to work in the University of Berkeley in California.

When the war ended, Wilkins had some domestic issues to iron out. His American girlfriend, Ruth, an art student, whom he has met in California, was pregnant and he suggested that they get married and keep the child. He intended to take up an offer of work with his old boss, J.T Randall, in St Andrew's University in Scotland after the war ended. He told Ruth that it rained a lot there. He later recalled that she replied, 'I like rain.' The couple got married and a son arrived but inside a few months Ruth wished to end the marriage. In June 1945 Wilkins returned to Britain alone, aged twenty-eight, heartbroken and divorced. It was a turbulent time for him and he was unsure which direction his scientific career should take. He read Erwin Schrödinger's book *What Is Life?*, based on a series of lectures Schrödinger had given in TCD in early 1943 (*see* Chapter 15). This was the first time a physicist of the highest order had addressed the biggest question for biology. Schrödinger looked at the problem from the point of view of a physicist, outlining the characteristics he believed the material that held the genetic code must have. The book had a huge influence on Wilkins's decision to abandon physics and join J.T. Randall's new biophysics lab in Scotland.

The book also encouraged a number of other young scientists who might otherwise have become physicists to get involved in biological research: these included Watson and Crick. The book was a call to arms for a generation of physicists, to apply their analytical skills to finding the unique code for life.

Despite the inspirational effect that reading *What is Life* had on Wilkins, he hit a low point shortly thereafter. The true implications of his work on the Manhattan project were laid out in shocking detail when he read a newspaper headline: 'Japanese City Destroyed'. The atomic bomb had been dropped on Hiroshima on 6 August 1945 and he had contributed to it. Wilkins felt small and ashamed. His mind was made up – biophysics it was.

St Andrew's was not an ideal place for Wilkins in 1945. He was in his twenties, single again and keen to socialise and get out and about. He began to feel that he needed to be more at the centre of biophysics. He had thought he would go to Scotland as a married man with a young child, but that dream had fallen apart. Furthermore he was having rows with his boss, J.T. Randall. He was looking for a new opportunity when Randall was made Head of Physics at King's College in London. Despite the arguments they had been having, he decided to go with Randall to the big city and make a new start.

Randall had funding from the Medical Research Council to set up a biophysics research unit and he began to build a talented team of researchers around him. After a while, he asked Wilkins to investigate how DNA – one of the molecules on the suspect list for being the material that held the code for life – moved about and grew in cells. Wilkins saw this as an exciting new field and gladly accepted Randall's suggestion. He became the assistant director of the Biophysics Research Unit in King's College.

Then came the big news from Oswald Avery's lab in Rockefeller University in New York: proof that DNA was the carrier of genes, the ephemeral elements that dictate a person's physical features and influence their personality traits. Avery had clearly shown that DNA was the vehicle by which genes moved around inside cells and between the generations. It was now obvious that anyone interested in finding an answer to Schrödinger's big question, 'What's is Life?' must focus on DNA.

That's what Wilkins did, using microscopes to look at cells. In 1950 he had a stroke of luck when he attended the talk by Rudolf Signer in Cambridge and obtained a phial of the highest quality, the most purified DNA in the world. How should he use it?

Wilkins recalled that J.D. Bernal had used something called x-ray diffraction to study living viruses. The idea was that x-rays fired at the viruses would be deflected as they bounced off its molecular components. The pattern of deflection, or diffraction, enabled Bernal to work backwards and reconstruct the structure of the virus. The approach is something akin to what a military power does when they capture a piece of technology belonging to the enemy. They pull it apart, examine the pieces and put it together again. Wilkins knew from this work that it was possible to use x-ray techniques to find the structure of living tissue.

Could x-ray crystallography be used to determine the structure of DNA? If so, this would be a major step forward for science, as the structure was still unknown. It might also reveal clues to how DNA worked, how genes were passed from one generation to the next, in fact rather than just theory. In science, particularly biology, researchers know that determining the exact physical structure of an organic entity goes a long way towards revealing how that entity works.

Wilkins and one of Randall's students, Raymond Gosling, set about examining the Signer DNA using x-ray diffraction. Using a watchmaker's tweezers, Wilkins managed to pull apart a dozen individual fibres of the DNA and line them up in parallel across a wire frame for analysis. Gosling had already tried unsuccessfully to get clear x-ray patterns, using DNA from sperm heads, but this time the results were stunning, the patterns much sharper and more detailed than ever before. This was important because the sharper the diffraction pattern, the more regular a structure will be. The result implied that DNA had a very regular structure, something that Wilkins and Gosling considered a prerequisite for any material that was the code for life. Whereas Avery had evidence that DNA carried the genes, Schrödinger and others suggested that DNA itself *was* the genes. The photograph, which was a series of tiny dots lined in an x shape, set the stage for what was to come.

Later that summer of 1950, Randall sent Wilkins to a scientific meeting in Naples, where he presented his and Gosling's photograph, the first clear evidence that DNA had a crystalline, or ordered structure. At the meeting in Naples, he met Jim Watson, who was very excited about Wilkins's photograph because it meant that the structure of DNA could now be established. Watson was determined to find that structure.

The DNA work at Randall's lab was making exciting progress and it was decided to bring in an expert in x-ray techniques to enhance the team. The person recruited was Rosalind Franklin, a talented and ambitious young researcher. But, as things turned out, Wilkins and Franklin found it impossible to communicate and their working relationship broke down completely. As Wilkins told it, Franklin was difficult, prone to outbursts and creating terrifying scenes. Others have given a different version of events but it is beyond question that this breakdown in relations was damaging for Randall's lab as a whole, as it meant that neither Wilkins nor Franklin really knew what the other was working on. Over time, Wilkins came to believe that DNA had a helical structure (which was later proved correct), while Franklin, at least initially, felt it was non-helical. The lack of cooperation between Wilkins and Franklin enabled other teams of researchers to catch up with them, notably the young and hungry Crick and Watson team in the University of Cambridge.

The presentation of the photograph at the Naples meeting effectively sounded the starting gun for the race to detail the precise structure of DNA and reveal its function. Less than three years after seeing the photograph in Naples, Watson and Crick overtook Randall's team and announced to the world that they had solved the puzzle of DNA – that it was composed of two tightly wrapped helical strands, connected by weak hydrogen bonds. The big prize had slipped through the fingers of the feuding Wilkins and Franklin.

The following story illustrates how close the Randall lab came to making the biological discovery of the century. One day in January 1953, as Franklin was preparing to leave for pastures new, Gosling, with whom she had been working and who had formerly worked with Wilkins, approached Wilkins with a photograph. It had been taken by Rosalind some time in 1952 and showed clear evidence that DNA was a double helix. This was the missing key, as Wilkins had assumed that DNA was made of three strands, not two. In his excitement, Wilkins showed the photo to Watson on one of the visits Watson made to the Randall lab. The image had a very distinct x shape that strongly indicated the molecule was helical. With hindsight, Wilkins may have made a strategic mistake.

After this, Watson and Crick, now more than ever convinced by the second key photo to come out of the Randall lab to indicate that DNA was helical in shape, launched into an all-out effort to build a model of what the structure of DNA might look like, using wires and balls. Three months later, in April 1953, they published a short paper announcing to the world that DNA was a double helix and that there was a big chance that it was the material with the magic code. However, Watson was anxious to acknowledge Wilkins's contribution to the discovery that made headlines around the world and invited him to be a joint author on the follow-up paper he and Crick wrote for *Nature*. Wilkins declined but published his own paper, detailing his work in generating the original Naples photo. Franklin, too, published her own DNA paper. In this way, the two main researchers of the dysfunctional Randall team published separately after the DNA horse had bolted. The fact that Watson and Crick got there first must have been a bitter pill to swallow as Wilkins and Franklin had showed them the way.

There was still important scientific work to be done. The structure of DNA, as outlined by Crick and Watson, had to be confirmed in the lab. Wilkins led the charge here and spent much of the next decade providing experimental proof that Crick and Watson's model was correct, for the most part. Sadly, Franklin, who was also central to the DNA story, died of breast cancer in 1958, aged just thirty-seven, missing out on a likely share of the 1962 Nobel Prize for Medicine and Physiology (a prize that can be given only to the living) that was awarded to Watson, Crick and Wilkins

The prize was to prove the high point of Wilkins's scientific career but he was still only forty-six and had a lot of living to do. He married for the second time in 1959, a woman named Patricia Chidgey, who had caught his eye one evening in the Institute for Contemporary Arts in London. Wilkins described Patricia as a lively, beautiful person interested in dancing and the arts, as he was. The couple had four children together.

Wilkins had been interested in anti-war politics from his student days in Cambridge in the 1930s. In the 1960s this interest was reawakened as he experienced the existential threat to mankind from the Cold War, as best exemplified in the Cuban Missile Crisis of 1962. He was an outspoken opponent of Britain placing US ballistic missiles in RAF Greenham Common in the early 1980s and he contributed to debates about the impact of science on the wider society.

Wilkins continued to do research at King's College up to his death in 2004. However, it appears that the breakdown in his relationship with Franklin haunted him for the rest of his life. Many popular books were written about the discovery of DNA and Wilkins felt that some of these unfairly depicted him as a misogynist and a bully in his dealings with Franklin. His desire to put his side of the story into the public domain was one of the main reasons he wrote his autobiography.

Wilkins probably regretted that things had not been better between himself and Franklin, as they could have solved the DNA puzzle before Watson and Crick. After Franklin's death, one of her personal notebooks surfaced, which showed that she had been thinking about DNA as a double helix with a spiral of 'three-eighths separation' (which was correct) at least a month before Watson and Crick produced their double helix DNA model. Wilkins had separately developed clear ideas about how the two DNA strands were bonded. If they had shared their knowledge, the names Wilkins and Franklin rather than those of Watson and Crick might forever be associated with the discovery of DNA. In his autobiography Wilkins admitted, 'If she and I had discussed the problem [the structure of DNA] there would have been little to prevent us finding the double helix.'

The Legacy of Maurice Wilkins

- He played a key role in the identification of DNA as the genetic material.
- He confirmed the double helix structure of DNA using x-ray photography.
- He participated in the Manhattan project, which led to the development of nuclear weapons.
- He was one of a pioneering group of physicists who devoted themselves to biology after the Second World War.
- He was a prominent anti-war activist in post-war Britain.

17

IT'S ONLY LOGICAL:

George Boole (1815-64)

Birthplace: born in Lincoln but spent his entire academic career in Cork

There are between 6500 and 7000 languages spoken across the world today, most of which use of an enormous variety of letters, symbols and words. In English, for example, we have 26 letters, which can be combined to make up more than a million different words. The same thing can therefore be said in English – and in many languages – in a multiplicity of ways. This makes for interesting conversation, the production of great works of literature and engaging oratory but it also means that billions of people, from Timbuktu to Tipperary, are, every day of their lives, using far more words when speaking and writing than is necessary for them to say what they need to say.

With the advent of electronic devices a more direct system of communication, one that would function between machines, was needed. Ideally this would be a simple system, facilitating completely waffle-free communication. In the 1930s, Claude Shannon, an Irish-American mathematician, realised that something had already been invented that could underpin a new machine language. This was Boolean logic, which was created by George Boole, a self-taught man from a modest background, and the first Professor of Mathematics in Queen's University Cork in 1849.

Shannon's 20th-century rediscovery of Boole and the application of his logic to electronic circuits and later electronic devices has had an enormous impact on our modern word. It could be said that the most common language in the world today is that 'spoken' by electronic devices, talking to one another using principles invented by George Boole. This means that anytime anyone anywhere uses a mobile phone, laptop or CD player, Boole's ghost hovers above.

These days Boolean logic is intimately associated with computers and electronic devices of all kinds. However, when it was invented, just a few years after the end of the Great Famine, the idea of a universal, electronic computing machine was way beyond the conception of even the most intelligent minds of the day. Boolean logic arose out of Boole's purely academic interest in devising a way to express clearly and simply, in algebraic terms, how the human mind makes its decisions based on certain logical assumptions and probabilities. This interest eventually led to the creation of his masterpiece, *An Investigation into the Laws of Thought on Which Are Founded the Mathematical Theories of Logic and Probabilities*, written while he was living at 5 Grenville Place, Cork, beside the River Lee, in the early 1850s. Though no one knew it at the time, this book, which described a new logic – Boolean logic – would become one of the most important ever written, utterly changing the world.

The man who became a central figure in laying the foundations for the world of information technology in which we live today, with its ubiquitous iPhones, tablets and desk-top computers, was born into a humble family in Lincoln in the east of England. His father was a shoemaker and, although intelligent and able, never achieved anything of note in his life. From a very early age, Boole displayed an extraordinary natural intellect. He taught himself French and German and with a little help from a tutor he became proficient in Latin. He never attended a university or college. In the light of this scant early education, his later achievements were miraculous.

Boole's reputation for brilliance grew quickly and early in his life in the small town of Lincoln and its surrounds. There was talk locally of his becoming the next Isaac Newton, who had also been born in Lincolnshire, in the town of Woolsthorpe-by-Colsterworth. When a public meeting was organised in Lincoln to honour Newton, the person asked to deliver a talk about the great man's legacy was nineteen-year-old George Boole. He was the logical choice of speaker, which indicates the high regard it which he was already held. Although initially reluctant, he reportedly delivered a lecture of confidence and authority and even had the audacity to suggest where Newton had gone wrong in some of his researches.

In Ireland or Britain today such an outrageously gifted individual as Boole, even if born into a poor family, would expect to finish secondary school and find his way to university by means of a scholarship. However the circumstances of the time and Boole's own family situation militated against this possibility. When Boole was in his mid-teens, his father became ill and could no longer work so he became the sole breadwinner of the family, supporting his parents and three younger sisters. This remained the case for most of his life. Some students of Boole's life believe that this financial pressure and early responsibility contributed to his bouts of ill-health later on, occasional low moods and frustration and perhaps even his premature death at the age of forty-nine. But he was able and resilient enough to cope most of the time and got on with his life with the constraints that providence had dealt him.

Boole's response to adversity was impressive. At the age of sixteen, he landed a teaching job at Heigham School in nearby Doncaster. He taught there for a few years, then, wishing to work closer to home, set up his own school for day pupils in Lincoln at the age of nineteen. This may seem remarkable today but for Boole and the people of Lincoln it was deemed a natural move. He already had a reputation in his hometown for genius and good character, so parents in the town and its surrounds were happy to send their children to his school. He was also involved with other schools, taught at a boarding school in Lincoln and later set up a school of his own in Doncaster. The work proved all-consuming. Although he recruited some members of his family in to help, ultimately he was responsible for the day-to-day running of the school, all the administration and the finances and dealing with the concerns of pupils and parents. On top of that he taught.

Boole spent a lot of the little free time he had at the Lincoln Mechanics Institute. Institutes like this existed across 19th-century Britain, set up primarily to cater for working-class adult men and often supported financially by wealthy local industrialists. The reasoning behind them was that wealthy patrons felt that it was better that working men 'upskilled' and became more knowledgeable, particularly in technical subjects, by reading technical and scientific books available in the local institute libraries, than that they spend their free time idly drinking in pubs.

At the Lincoln Institute, Boole read texts such as Newton's challenging masterpiece, *Principia,* and the works of the great French mathematicians Lagrange and Lacroix. He ploughed through difficult text after difficult text and grew thirsty for more. As his knowledge grew, he campaigned for the extension of the institute's reading list to include more advanced and important mathematical and scientific texts. He read these books slowly and carefully as he had little or no knowledge of the assumptions or prior information upon which they were based. Nevertheless, he made steady progress.

During this period, Boole largely accepted the family burden on his shoulders without complaint. In letters he wrote in later years, however, he did express frustration that he couldn't afford a private tutor and had little time to pursue his growing interest in maths, science and technical subjects. This meant it took a lot longer for him to reach his potential than would otherwise have been the case. But this slow slog towards his ultimate destination might not have been totally devoid of benefits: it meant that he developed a powerful capacity for independent creative thought and for developing his own insights. This talent, some suggest, might have been blunted had he followed the usual path of talented mathematicians, then and now, into a university.

By the time he had reached his early twenties, Boole had completed an extensive programme of self-taught, high-level study. He had been advancing his knowledge methodically, taking notes and making observations. He began to come up with ideas on how to advance even the findings of some of the great masters. He felt he had a contribution to make and this conviction grew to the point where he considered submitting a paper detailing his thoughts to a professional scientific journal.

In 1839, at the age of twenty-three, he took the plunge and sent a paper for publication to a specialist mathematics journal. He risked rejection, of course, and the fact that he was unknown in the small British and Irish mathematics community at the time meant he had a credibility barrier to overcome. It was doubly fortunate for Boole that the *Cambridge Mathematical Journal* had just been founded with the sole purpose of publishing original research and that its editor was a far-seeing young man called Duncan F. Gregory, a graduate of TCD and a talented mathematician in his own right.

Gregory immediately recognised the potential of the paper by the unknown Boole when it landed on his desk but rather than publishing it immediately, he became something of a mentor to Boole and spent the next two years or so advising him on how to improve his writing and the presentation of his work so that it was more acceptable for academic publication. At the same time he encouraged Boole, helping to improve his self-confidence. Gregory's kindly mentoring was crucial in bringing Boole's talent to fruition.

In 1841, Gregory finally published Boole's first paper. This was a key moment as it meant that, at least within the small circle of top mathematicians, his name became known. He was no longer a complete outsider, although there was always a small number of leading mathematicians who didn't like to see a working-class man gain admittance to their elite gentleman's club. The surge of adrenaline Boole got from seeing his name in print encouraged him to even greater efforts and over the next few years, Gregory published other papers by him. By 1844, he believed that he had come up with something that deserved to be published in a journal of even higher repute than the *Cambridge Mathematical Journal.* Showing remarkable courage, Boole set his sights on one of the journals produced by the Royal Society, the premier scientific society in Britain and one of the most prestigious such societies anywhere in the world. It was a particularly ambitious move because the Society generally published only the work of the most famous and well established researchers and did not publish many papers about maths.

It was perhaps inevitable that significant opposition to the publication of Boole's paper arose inside the committee of the Society that was responsible for deciding whether a paper should be published or not. This opposition was apparently based almost solely on the fact that the author was not from the expected social background. The story goes that after a heated argument in committee, it was decided to send Boole's paper for consideration to an independent expert, Philip Kelland, Professor of Mathematics in the University of Edinburgh. Kelland recommended not only that it be published but that its author should receive a special prize. Boole was duly awarded the Society's new Gold Medal for Mathematics in 1844, a great honour that opened up career possibilities for him outside teaching.

The time had finally come, Boole believed, to start following his true passion – mathematics – and he began to explore whether he could get a university job. An opportunity arose in 1846 when three new Queen's Colleges were set up in Ireland, in Belfast, Galway and Cork. Boole was encouraged to apply for a professorship at one of the new Irish colleges by his growing band of academic supporters, including the influential Belfast-born William Thomson, later Lord Kelvin [*see* Chapter 5]. He assembled an impressive group of people who offered testimonials on his behalf and formally applied for 'a Professorship of Mathematics or Natural Philosophy [Physics] in any of Her Majesty's colleges now in the course of being established in Ireland' in a letter sent from Lincoln, dated 15 September 1846. Boole hoped that the quality and the range of people supporting his application would compensate for his lack of formal education: distinguished academics like Augustus De Morgan, Professor of Mathematics at the University of London, Philip Kelland from the University of Edinburgh, the Mayor of Lincoln and some of its other leading citizens. He also had important local support from the Reverend Charles Graves, Professor of Mathematics in TCD.

The process of appointing new professors in Ireland was agonisingly slow and dragged on through 1847 into 1848. At one point, Boole became so frustrated with the delays that he withdrew his application altogether but he changed his mind and applied again. On 12 December 1848, his father died. This was a very sad event for Boole but in practical terms it meant that he had one less person to support. Finally, in August 1849, just one month shy of three years after he applied for the job, he received the news that he was to be appointed the first Professor of

Mathematics in Queen's College, Cork. He was thirty-three and at last he could begin his life as a professional mathematician.

Boole initially found it difficult to settle in Cork, the city where he would spend the rest of his life. In letters of this period, he oscillates between saying he was happy in Cork and saying he greatly missed Lincoln. He also found himself embroiled in a number of serious in-house squabbles in the university that were stressful and time-consuming. However, he soon became a highly respected figure among the students as well as most of the staff.

In his mid-thirties, he first met his future wife, Mary Everest, the favourite niece of Sir George Everest, when she made a visit to Cork in 1850. (The Royal Geographical Society named Mount Everest in his honour in 1865.) Mary was just eighteen and was visiting her uncle, John Ryall, Professor of Greek. The young woman, who was interested in maths and later became a feminist philosopher and educationalist, was introduced to Boole and they became close friends. In 1855, Mary's father, Thomas, died, leaving her nothing in his will. Boole – always a man of honour – stepped in and proposed marriage to Mary, apparently an act of love as much as one of charity.

The years just before his marriage, from 1852 to 1854, were when Boole composed his masterpiece, *The Laws of Thought.* He invented Boolean logic as he looked out at the Lee from the desk of his study in Grenville Place. He created a new mathematical language that was based on two symbols, zero and one, choosing just two symbols because a language could be built up around this 'binary' system. A century or so later Boolean would prove to be the perfect means by which electronic devices and computers could 'talk'. The potential of Boolean logic to underpin communication across electronic devices was first recognised by Claude Shannon – regarded as 'the Father of Information Technology' – after hearing about Boole's work at a philosophy class he attended.

In 1937, Shannon wrote a master's thesis in Massachusets Institute of Technology (MIT) showing how Boolean algebra could improve the design of systems then used as telephone routing switches. This insight led to Boolean being used by electronic engineers when they came to design computers and other advanced electronic devices from the 1960s on. 'Boolean', with its utter lack of redundancy, was, Shannon believed, the ideal language to control electronic circuits.

Information theory, as outlined by Shannon, who conceived it, concerns the study of how much information it is necessary to include in a message to get the message across. A good example of this is texting language. When people compose text messages, they do it in different ways. Some people type in all the letters of all the words, with the right punctuation and grammar. Such grammatically perfect messages are also high in redundancy. We all know some shortcuts, such as 'l8r' for 'later' or 'omg' for 'oh my God'. But, how far can we pare this down? How small can the message become and still retain its essential, real information? Shannon came up with a mathematical way to measure this and in the process he rediscovered Boole.

The beauty of Boolean, for Shannon, was that it was a language built around just two 'letters' or two states of existence. This was ideal for controlling an electronic circuit, as a circuit can exist in only one of two states – on or off. It is an undeniable fact that electronic circuits are made up of wires and either there is current running live through the wire, or there isn't. However, endless communication possibilities can be formulated through combinations of ons and offs, the two Boolean letters.

From the middle of the 20th century, increasingly complex electronic circuits were controlled and operated by Boolean logic. As the computer age emerged in the 1960s and 1970s, engineers came up with the idea of 'gates' within a circuit. They built something called an Xr gate, where there were two wires going into a circuit and one coming out. The wire going out was said to be the Xr of the two wires going in. If both incoming wires are off or on, then the Xr is off. If one of the incoming wires is on and the other is off, then the Xr is on. The Xr is, therefore, the product of the interaction of wires and the on and off states inside a circuit.

There are other possible manipulations of on and off, which help to add complexity and operational capacity to an electronic circuit. For instance, there is something called *negation*, with one wire going into a circuit and one going out. The wire coming out is always in the opposite state to the one going in. A third combination routinely used in circuits is called AND by computer scientists. This is where the output wire is on only when both wires going in are also on. All these combinations can be put together in various ways to make up a complicated circuit. A modern microprocessor, a computer's brain, will have millions of these different types of gates.

Boolean is everywhere today, says Professor Gary McGuire, Director of the Claude Shannon Institute at UCD and a world renowned mathematics and encryption expert. 'Boole's impact on the modern world has been enormous. All our electronics, our mobile phones, transistors, circuits, microchips are based on Boolean logic. You can't get any simpler than two symbols – everything is reduced to zeros and ones. That is ultimately what computers do, they reduce everything to zeros and ones. It is as if we were communicating using just the letters a and b. There will be the same variety; it is just that the 'words' will be longer. Anything can be built up from the basic 'gates'. Boole's alphabet is zero and one. In English there are twenty-six letters. Break language down to its simplest terms and it is made up of two symbols, like Morse Code dots and dashes.'

When Boole invented his Boolean he had no idea that it would be used to underpin the global communications edifice of 21st-century society and commerce. However, he did have a good idea that he had come up with something that helped to better define how the human mind works and how it makes its logical decisions. There had been many attempts to define the process of human reasoning: Boole, in *The Laws of Thought*, showed that much of it was based on the balancing of possibilities and probabilities. A new algebra was created and the fields of logic and probability were reinvigorated.

The Laws of Thought had a huge impact almost immediately and it has remained in print ever since it appeared in 1854. Boole had become successful, he had a Royal Society medal, many publications and two books under his belt and he was not yet forty. Life was good and he reached a level of contentment that had not been possible earlier in his life. He was also fortunate to have a very happy marriage and a supportive wife. The only dark cloud was that he suffered from a lung complaint and rheumatism, conditions not helped by living in the damp climate of Cork. In 1856, the couple had their first daughter, Mary, and Boole was reportedly in a state of delight following the arrival of his little 'beauty'. They went on to have four other daughters – Ethel, Alicia, Lucy and Margaret. Alicia made promising discoveries in the field of four-dimensional geometry as a self-taught mathematician like her father; Lucy became the first female professor of chemistry in Britain; and Ethel became a supporter of revolutionary causes and author (under her married name of Voynich) of *The Gadfly* (1897), a novel that sold two and a half million copies in the Soviet Union.

The Booles lived for a while in College View on Sunday's Well Road but after the arrival of their fourth daughter, Lucy, in 1862, the family moved to the picturesque Lichfield Cottage in Ballintemple. This was big enough to accommodate the growing family and they were able to employ household help. Boole walked to the station in Blackrock every day and took the train to work.

His students greatly respected him and were amused when, like many professors, he became lost in thought, even during a lecture, pondering some difficult mathematical problem. At home, he was a loving, although strict father. He was a man of the people in the sense that he would easily strike up a conversation on any subject with anyone interested. He would even invite people he had just met over to his house to look at his telescope or microscope, if they had expressed an interest in science or astronomy. He was an emotional, loyal and sensitive man, totally committed to his work and to his wife and daughters.

Boole shared with his wife Mary an interest in what today would be called homeopathic medicine. He was prepared to seek alternatives to conventional medicine but in politics he was conservative. He was not a supporter of reform, or of improving the conditions of the working poor, yet he was very kind and generous to his students. He had a love of animals, yet appeared to be blind to the horrific human suffering in rural parts of Cork in the aftermath of the Famine.

On 24 November 1864, Boole set out for work. It was raining heavily and apparently he didn't attempt to get the train, as he feared he would be late for a lecture. Instead, he decided to walk the three miles to College in the rain. He got completely saturated and later that evening, when he got home, he began to feel ill and feverish. One story has it that his wife Mary adopted a homeopathic approach to trying to cure her husband of his fever, by constantly pouring water on his body and making him worse. In any case, his condition steadily worsened over the next few weeks and he died on 8 December 1864, at the age of forty-nine. The certified cause of death was pleura-pneumonia. He was buried in St Michael's Church of Ireland in Blackrock, Cork.

Boole had striven for years to reach his goal of becoming a professor of mathematics, so it was sad that his life was cut short before his career had reached its natural end. On the family front, he witnessed the birth of his five girls but did not live long enough to watch them grow into adults and make their mark on the world. He also died without knowing what a huge impact his work would have.

Meanwhile the house in 5 Grenville Place, Cork, in which Boole invented the language of computers, lies derelict and its future is uncertain. Surely, someone should step in to preserve this house where the foundations of our modern world of instant electronic communications were laid, in memory of a man who shaped the course of human history.

The Legacy of George Boole

- He invented Boolean algebra, the language of all electronic devices.
- He was influential in the mathematical fields of logic and probability.
- He was the first Professor of Mathematics in Queen's University, Cork (now University College, Cork).

BIBLIOGRAPHY

Books

Cosslett, Tess (ed.). *Science and Religion in the Nineteenth Century.* Cambridge University Press, 1984.

Courtney, Nicholas. *Gale Force 10: The Life and Legacy of Admiral Beaufort.* Headline Book Publishing, 2002.

Friendly, Alfred. *Beaufort of the Admiralty: The Life of Sir Francis Beaufort 1774-1857.* Random House, 1977.

MacHale, Desmond. *George Boole: His Life and Work.* Boole Press Limited, 1985.

McLoughlin, P.J. *Nicholas Callan: Priest-Scientist (1799-1864).* Clonmore & Reynolds, 1965.

Morris, Richard Knowles. *John P. Holland 1841-1914: Inventor of the Modern Submarine.* University of South Carolina Press, 1966.

Mulvihill, Mary (ed.). *Lab Coats and Lace: The Lives and Legacies of Inspiring Irish Women Scientists and Pioneers.* Women in Science & Technology, 2009.

Ray, Andrew. *Lord Kelvin: An Account of his Scientific Life and Work.* J.M. Dent & Co, 1908.

Sexton, Michael. *Marconi: The Irish Connection.* Four Courts Press, 2005.

Schrödinger, Erwin. *What Is Life?* Cambridge University Press, 1944.

Soon, Willie Wei-Hock and Steven H. Yaskell. *The Maunder Minimum and the Variable Sun-Earth Connection.* World Scientific Publishing Co. 2003.

Thompson, Silvanus P. *The Life of Lord Kelvin.* AMS Chelsea Publishing, 1910.

Wallace, W. J. R. *Faithful to Our Trust: A History of the Erasmus Smith Trust and the High School, Dublin.* The Columba Press, 2004.

Wilkins, Maurice. *The Third Man of the Double Helix: An Autobiography.* Oxford University Press, 2003.

Other Publications

Brock, W. M. 'John Tyndall', *Oxford Dictionary of National Biography*, 2004.

Bruck M.T. and S. Grew. 'The Family Background of Annie S.D. Maunder (née Russell).' *Irish Astr.* J. 23 (1), pp. 55-56 (1996).

Bruck, M.T. 'Alice Everett and Annie Russell Maunder: Torch-Bearing Women Astronomers.' *Irish Astr.* J. 21 (3/4), p. 281 (1994).

Burns, Duncan T. Robert Boyle Science Festival 2011, Lismore Lecture. November 2011.

Casey, M.T. 'Nicholas Callan, Priest, Professor and Scientist.' *Physics Education*, Vol. 17, 1982, pp. 224-34.

Callan, Rev N. 'On the Induction Apparatus.' *Philosophical Magazine*, Nov. 1857.

Doyle, Major D. J. A.M.S. 'The Holland Submarine.' *An Cosantóir, Irish Defence Journal*, Vol. VII, No. 6, June 1947.

Holland, John P. 'The Submarine Boat and Its Future.' *North American Review*, December 1900.

Kennedy, Peter. 'Marconi and Ireland.' A commemorative lecture on the centenary of Marconi's Nobel Prize for Engineering. UCC, 2009.

Manchester, Richard ('Dick'). Address to the Pulsar Conference in Sardinia, 2010.

Reville, William. 'John Tyndall', *The Irish Times*. 5 April 2001.

Rose, Steven. 'Leaders of National Life: Professor Maurice Wilkins FRS.' British Library, 1990.

Spearman, David. 'William Rowan Hamilton'. *Dictionary of Irish Biography*. Royal Irish Academy, 2011.

Struther Arnott, T.W.B. Kibble and Tim Shallice. 'Maurice Hugh Frederick Wilkins, Elected FRS 1959.' Biogr. Mems Fell. R. Soc. 2006, p. 52.

'The Phoenix. (John P. Holland 1841-1914): The Liscannor Man Who Invented the Sub'. *Clare Champion*, 9 August 1996.

Tyndall, John. *Address delivered to the British Association Assembled in Belfast: with Additions* (1874). Longmans Green & Co., 1874.